KINETIC AND THERMODYNAMIC ASPECTS OF POLYMER STABILITY

KINETIC AND THERMODYNAMIC ASPECTS OF POLYMER STABILITY

G.E. ZAIKOV (EDITOR)

NOVA SCIENCE PUBLISHERS, INC.

COMMACK

Art Director: Maria Ester Hawrys
Assistant Director: Elenor Kallberg
Graphics: Denise Dieterich, Kerri Pfister,
        Erika Cassatti and Barbara Minerd
Manuscript Coordinator: Sharyn Schweidel
Book Production: Tammy Sauter, and Benjamin Fung
Circulation: Irene Kwartiroff and Annette Hellinger

Library of Congress Cataloging-in-Publication Data

Kinetic and thermodynamic aspects of polymer
stability / G.E. Zaikov (ed.).
        p. cm.
Includes bibliographical references and index.
ISBN 1-56072-258-4 :
        1. Polymers--Deterioration. I. Zaikov, Gennadii
Efremovich.
TP1122.K497 1995   95-24688
620.1'920421--dc20   CIP

© *1996 Nova Science Publishers, Inc.*
  6080 Jericho Turnpike, Suite 207
  Commack, New York 11725
  Tele. 516-499-3103 Fax 516-499-3146
  E Mail Novasci1@aol.com

Printed in the United States of America

CONTENTS

FIRE AND HEAT SHIELD MATERIALS BASED ON SULFOCHLORINATED POLYETHYLENE, 209

M.A. Shashkina*, R.M. Aseeva, A.A. Donskoy* and G.E.   Zaikov

KINETICS AND THERMODYNAMIC ASPECTS OF POLYMER STABILITY

G.E. Zaikov
Institute of Biochemical Physics, Russian Academy of Sciences
4, Kosygin Street, Moscow 117334, Russia

"Life degradation as well as the degradation of Society and State is bad, while Polymer Degradation is sometimes very well."

Professor N.M. Emanuel
(December 5, 1984)

INTRODUCTION

The present collection is volume of selected papers, which were written by scientists from former Soviet Union (sometimes in cooperation with scientists from Western Europe). The papers deal with fundamental research in the area of aging and stabilization of polymers, prediction of polymeric materials stability under storage and exploitation, combustion of polymers and fire retardancy, acceleration of degradation. The problems of pyrolysis, thermal and thermo-oxidative degradation, photodegradation, radiation-induced degradation, hydrolysis and mechanodegradation are discussed in details. Part of papers concerns with polymer treatment, transport phenomena in polymers, controlled release of drugs from polymer matrix. The aim of this spin-off is to inform English-speaking scientists about the last achievements of scientists from former USSR in the aforesaid areas of Chemistry and Physics of Polymers.

I would like to finish my introduction with a typical Russian joke.
Mr. Peter met Mr. Ivan who drove the bicycle.
Peter: Where do you go my dear friend Ivan?
Ivan: I am going to the United Kingdom.
– For what reason?

– I would like to marry with HRM Elizabeth II.

– Is it possible? Mr. Sidor told me that She is already married with HRH Duke of Edinburgh.

– This problem is half-solved at present because as I am concerned I am ready, but I did not ask (yet!) opinion of Her Majesty.

Apparently, we follow Ivan's idea. We also did not ask (yet!) opinion of readers, but we would like that our book would be useful for scientists, who are experts in the areas of polymer chemistry, application of polymer for industry and medicine as well as for experts in the area of chemical and biological kinetics.

We will be very happy to receive some comments and recommendations about this volume which will try to use in further research and selection of papers for publication in the next volume.

Professor Guennadii E. Zaikov,
Institute of Biochemical Physics,
Russian Academy of Sciences

BASIS OF THE KINETICS
OF CHEMICAL PROCESSES

E.F. Vainshtein and G.E. Zaikov

Institute of Chemical Physics, Russian Academy of Sciences,
4 Kosygin Str., Moscow 117334, Russia

Abstract—The fundamental approach to the description of the kinetics of chemical reactions has been considered, and principle of chemical reaction independence and active mass law have been taken into account. Models for rate constants of elementary reaction have been discussed on the basis of the collision theory, theory of activated complex and the theory which takes into account the movement of intermediate particle to the reaction barrier.

The collision theory is suggested to use at high temperatures, the theory of activated complex is recommended to apply at moderate temperatures, and the theory, which takes into account the movement of intermediate particle to the reaction barrier, is relevant to low temperatures.

It is emphasized that the principle of maximal change of free energy, which was suggested earlier, may be applied for analysis of the mechanism of chemical processes.

As it has been shown in the previous paper [4], chemical reaction may be performed in the system in presence of thermodynamic opportunities. The description of this reaction is performed according to the methods of chemical kinetics. Evidently, degradation processes, as well as any others, must be described using both kinetic and thermodynamic parameters. However, this connection is displayed insufficiently for the most number of the known processes.

As a rule [2,3] chemical reactions represent the collection of elementary chemical acts, proceeding in parallel or regularly. It is supposed usually, that all chemical acts proceeds independently (the principle of chemical reaction independence).

The rate of the proceeding of each act (W) is described by the following differential equation:

$$W = \frac{d(m/V)}{dt} = kc_1^a c_2^b c_3^d, \tag{1}$$

where m - substance amount in moles, forming as a result of particular chemical act; V - system volume; k - rate constant of the process; c_1, c_2, c_3 - concentration of statistically independent particles, participating in the act, which characterize the reaction order by each component; a + b + d≤3 characterize total order of the reaction, defined by general number of colliding statistically independent particles. Proceeding in liquid and solid media, the reaction usually possesses the order not higher than the second. One of the reaction components is a polymer in the describing processes. In the most number of cases it is accepted, that monomeric unit is called as statistically independent particle [4,5]. Then in any case the rate of acts, proceeding during particular stage, can be expressed by the following equations:

$$W = kc_1 c_2 (a); W = kc_1 (b); W = kc_1^2 (c) \tag{2}$$

where c_1 - concentration of nonreacted active monomeric units of the chain; (a) relates to the reaction of the first type:

$$\sim\sim\sim\sim\sim +S \rightarrow$$

(b) relates to the reaction of the second type:

(c) - to the reaction of the third type:

where S low molecular substance (in rare case statistically independent parts of the chain molecule).

The application of the concentration of monomeric units for the description of kinetics according to acting masses law (equation 1) is based on the following statement of chemical kinetics: "the rate of the reaction by particular component of multicentral molecule is proportional to molecule concentration and the amount of active centers in it." In this case it is supposed, that all active centers possess equal reactionary ability. But if chain centers differ by reactionary ability, that is observed for the chains of limited size [6], then the rate constant k represents the mean value:

$$kn_1 = \sum_i k_i n_i,$$

where k_i - rate constants of the reaction of the active center i; n_i - the amount of i centers in the molecule; n_1 - total amount of nonreacted active centers. Evidently,

$$c_1 = n_1 N,$$

where N - concentration of polymer molecules.

If it is considered the formation of a new substance, then W>0. W<0 if the amount of the substance reduces. If the system volume is constant, that is observed in diluted solutions and gases, then m/V may be substituted by concentration. With regard to overall mentioned, the reaction of the chain break

$$\sim\sim\sim\sim\sim\sim\sim\rightarrow\sim\sim\sim+\sim\sim\sim\sim$$

possesses

$$-\frac{dc_1}{dt} = kc_1.$$

The reaction of chain fragment formation possesses

$$\frac{dL}{dt} = 2\,kc_1.$$

where L - concentration of fragments, being formed during degradation. Coefficient 2 points out, that two particles are formed during one act. But if the volume is variable, that is often observed in reactions, proceeding in solid phase, then [2]:

$$W = \left(\frac{1}{V}\right)\cdot\left(\frac{dm}{dt} - \frac{m}{v}\cdot\frac{d\ln V}{dt}\right). \tag{3}$$

Unfortunately, the change of system volume is calculated very rare in the process. For the reactions of the second order chemical acts consists with two consecutive stages: particle collisions, forming intermediate particle, and this particle transfer through the reaction barrier. In this connection life time of the particle is

$$\tau = \tau_c + \tau_d,$$

where τ_c - collision duration; τ_d - time of particle transfer over the reaction barrier. To describe reactions of the first order, it is required the time of particle motion over the reaction barrier only.

To understand physical sense of the process of the chemical act multiplicity of theories was suggested, [2,7,8] in any way characterizing the stages of chemical act. At present the most applicable are the theory of collisions and the theory of activated complex. As it follows from the name, the first theory describes the initial stage of the act under the supposition, that duration time of the second stage is much shorter, than of the first one. It is supposed in it [2,7], that reacting molecules, being considered as absolutely rigid ones, approach to the distance, equal half sum of their active diameters. At any active collision, i.e. such approach, the chemical act proceeds. It is accepted in the theory, that collisions do not break particle distribution, stipulated by their rates. That is why the rate constant of the reaction is estimated by the number of double collisions (Figure 1). The expression for the rate constant K of the reaction of the second order, obtained on the basis of the collision theory, is shown as follows:

$$K = pd^2 (8\pi kT / \mu)^{1/2} \exp(-E / RT) \frac{N_a}{1000} \, l / mole \cdot \sec. \qquad (4)$$

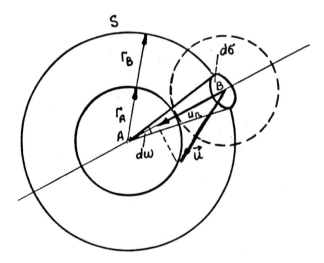

Figure 1. The collision of the center of particle B of area $d\sigma$ on the surface of the sphere S of radius $r_A + r_B$ and center A.

Here d - efficient diameter of colliding particles (molecules); kT - Boltzman factor; N_a - Avagadro number; $\mu = \dfrac{m_1 m_2}{(m_1 + m_2)}$ - reduced mass of colliding molecules, possessing masses m_1 and m_2; E - activation energy; p - steric factor, calculating .... orientation of particles in the space is required for the reaction proceeding. It is supposed, that exp (-E/RT) characterizes the part of active collisions, leading to the reaction, i.e. the part of collisions, in which molecule energy allows to overcome energetic barrier. The theory gives no methods of calculation of steric factor, which decreases with the growth of molecule size and their complication. In the most number of cases the calculation of it is performed, basing on the comparison of experimental and calculation data. The value of p is close to 1 for the reactions of very small molecules in solution, only. Apparently, p value depends weakly on temperature. At least, for relatively high temperatures this value is defined by probability laws of the collision of two reacting molecules, only; in this case it is supposed that at each moment of time any orientation of the molecule in the space is equally probable. The inclination to equal probability increases with temperature growth. The term exp (-E/RT) decreases also, approaching unit, with temperature growth.

As the reaction rate is defined by the number of active collisions, the role of exponential term in the expression for the rate constant of the reaction in the collision theory (equation 4) is not clear enough from physical point of view.

If it is accepted, that rebuilding of structure is connected with energetic barrier overcome, then the probability of collision must increase and the part of E/RT energy, necessary for the reaction barrier overcoming, must decrease with temperature growth. Apparently, energetic barrier overcoming takes relatively short time at relatively high temperatures only, i.e. the reaction rate is defined by the collision theory. Moreover, as it was mentioned above, the role of the collision theory is not clear enough in the description of the reactions of the first order.

The motion of particle, forming at the collision, over the reaction carrier (Figure 2) is considered in the theory of "activated complex" [2,3,9]. Initial configuration of atoms transits to the end one by means of continuous change of their relative disposition in any elementary chemical reaction. If the end and the initial state is relatively, then during the rebuilding their energy is higher, that in the initial state. According to thermodynamics free energy of system formation, in which direction chemical reaction must proceed, must be energetically lower. Naturally, the present statement relates to reactions, directed to the one side (irreversible). Evidently, the transition between initial and end mutual disposition of atoms is connected with the energy change. As the energy of initial and end configuration shows the minimum,

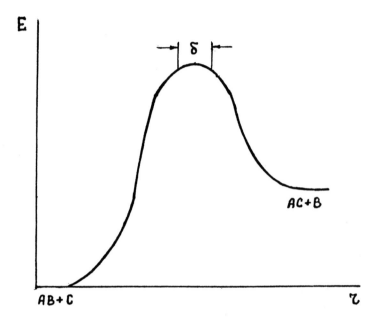

Figure 2. The change of potential energy E against coordinate of reaction r. δ is the area which corresponds to the transitional state.

there exists the conformation, in which initial configuration is destroyed incompletely, and end one is reached incompletely, and energetic maximum is reached in this configuration. This configuration, being critical, was named as "activated complex." It is clear, that if the system has reached this point, it reaches end state with particular probability. The function, describing the dependence of potential energy on mutual disposition of atoms, being rebuilt during the process, characterize the surface of potential energy. As the energy is multidimensional function of mutual disposition of atoms (distance and angle between atoms, number of atoms, etc.), then this energy dependence is multidimensional surface (each point on the surface is corresponded to a definite mutual disposition of atoms in transitional particle). Complete calculation of the function for various types of reactions was not performed successfully yet. That is why motion trajectory (mutual change of atom disposition) is not clear still. Evidently, it is impossible to describe such motion without additional simplifying suppositions. This surface is depicted by the method of horizontals, accepted in topography. In this case bulky surface is cut by equally distant horizontal plains, and cross section is projected to the basic plain. In this case it is possible the motion by different directions of potential energy surface. The way, most profitable from energetic point of view, is named as reaction coordinate. Consequently, reaction coordinate is the generalized characteristic of mutual disposition of atoms. The motion along the reaction coordinate is instable (it is connected with the energy increase). However, this way is mostly energetically profitable in relation to any other directions. Consequently, it is the most probable one. Let remind, that fluctuation probability is proportional to exp (- E/RT). In fact, the selection of the mostly probable way relates to the principle of the maximum of performed labor [10]. The most probable dependence, characterizing the change of potential energy with the reaction coordinate, is shown on Figure 2. The value, characterizing the difference between maximal energy ("activate complex") and the energy of the initial state is called real activation energy. Consequently, activate energy is the minimum excess of energy in relation to the initial state, which is required to chemical transformation. Considering the motion by the reaction coordinate it is supposed, that there occurs on break of equilibrium (Boltzman) distribution of particles by energies.

As any types of particles posses distribution by energies according to modern statements of physics, a part of them, possessing enough energy for reaching activated complex, with enter chemical reaction. It is clear, that the higher activation energy is, the smaller the number of particles is which are able to reach the state of activated complex. The reaction rate will be defined by the number of particles, reached activated complex, life time in it and the probability of particle lowering to the side of end state by the reaction coordinate. Life time of activated complex, i.e. the time of presence at the upper point of energetic barrier of the reaction, depends on its spread (δ). This part is considered as the one, parallel to the reaction ordinate axis. The parellelism of it is stipulated by the supposition, that there is no change of mutual disposition of atoms in the state of activated complex, and consequently, there is no energy change (dE/dr=0). That is why it is supposed, that the motion in activated complex is defined by the rate of kinetic motion, and, consequently, by the distribution of particles by the reaction rates (according to Maxwell). It is necessary to point out, that δ value is not expressed physically. Lifetime of activated complex is considered as the ratio of δ to the average rate of passing this part by activated complex. As the trajectory of forming particle is not clear, it is supposed, that there exists the equilibrium between initial and intermediate state. Such equilibrium differs from the one of chemical reaction, in which the rate of direct reaction equals the rate of reverse one. The present equilibrium means, that restoration of Boltzman-Maxwell distribution dye to the expense of activated complex proceeds much faster, than chemical transformation itself. Basing on the abovementioned suppositions, it was deduced the following equation for the rate constant of the reaction of the second order [2,9]:

$$K = \chi \frac{kT}{h} \cdot e^{-\Delta G^{++}/RT}, \tag{5}$$

where $\Delta G^{++}=G^{++}-G_A-G_B$ - the change of free energies of initial particles A and B during their transition into activated complex, kT - Boltzman factor, h - Plank constant, χ - transmissional coefficient.

There exists the correlation between activation enthalpy ΔH^{++} ($\Delta G^{++}=\Delta H^{++}-T\Delta S^{++}$) and experimental activation energy E_a:

$$\Delta H^{++}=E_a-RT+p\Delta V^{++}, \tag{6}$$

where ΔV^{++} - the change of the volume of particles at their transition to activated complex. For reactions in gas phase

$$p\Delta H^{++}\approx\Delta n^{++}RT,$$

where Δn^{++} - the change of particular number during the formation of activated complex. For the reaction of the second order $\Delta n^{++}=1$, and for reactions of the first order $\Delta n^{++}=0$. Then

$$K = \chi \frac{kT}{h} \cdot e^{\Delta S^{++}/R} \cdot e^2 RT \cdot \exp(-E_a/RT) \tag{7}$$

To deduce the equation (7) it was used the correlation for the change of entropy during the formation of activated complex under the supposition, that initial molecules and activated complex exist as ideal gas

$$\Delta S^{++}=\Delta S^{\circ}-\Delta n \cdot RT ln RT,$$

where ΔS° - standard entropy of a single mole of ideal gas.

For theoretical estimation of the rate constants of reactions it is necessary to make suppositions about the structure of activated complex, which studying by experimental methods is hard. That is why theoretical calculation of the rate constants of reactions is performed rather largely.

Comparing expressions for the rate constants in collision theory (equation 4) and in activated complex theory (equation 5) it is seen, that the latters possess constants, depending on both molecular and thermodynamic parameters. As it was mentioned above, one of the main advantages of the theory of "activated complex" is the possibility to apply it to the reactions of the first order, for which the rate constant equals

$$K = \chi \frac{kT}{h} \cdot \exp(\Delta S^{++} / R) \cdot \exp(-E_a RT) \tag{8}$$

It is probable that reaction description using the equation for the rate constant, based on the theory of activated complex, is more profitable at lower temperature range in comparison with the application of collision theory. Considering the reactions at much lower temperatures and the reactions of large molecules [11,12] it is difficult to suppose, that the time of particle motion over the reaction barrier may be identified with the life time of activated complex. For such reactions, in which the limiting stage of chemical act is the motion of particles, being formed at the collision of intermediate ones, over the reaction barrier, the following expression for the rate constant of elementary reaction was obtained [12]:

$$K = K_{eq} A \sum_{E_i = E_n}^{E_i = E_{max}} \exp(-E_i / RT) \lambda P_n / r_n \left(\frac{2m}{(E_m - E_o)\eta} \right)^{1/2} *$$

$$* \left[1 + \frac{r_k - r_n}{r_n} \left(\frac{E_n - E_o}{E_n - E_k} \right) \right] \tag{9}$$

A number of designations of the equation 9, namely E_n, E_k, E_o, r_o (0), r_k, r_n are shown of Figure 3. Here $A = e^{-G_o/RT}$, where G_o - free energy of the formation of intermediate particles in the initial state; K_{eq} - equilibrium constant of the formation of intermediate particles; P_k - probability that the particle existing on the top of the reaction barrier (in analog to activated complex) will approach the end state; λ - coefficient, calculating the probability for the particle, lowering from the barrier to the initial state, to come back to the top; η - coefficient of units recalculation; m - mass of intermediate particle or its part, which participate in elementary act (the latter relates to large molecules, for example to polymers).

The following model was used at the deduction of the equation (9). The motion of intermediate particles over the barrier was considered as mechanical process, as in the theory of activated complex. For such systems the rate of equilibrium setting of the formation of intermediate particles is much higher than the rate of chemical reaction proceeding. In this case the amount (concentration of intermediate particles, able to overcome the reaction barrier, equals the multiplication of transitional

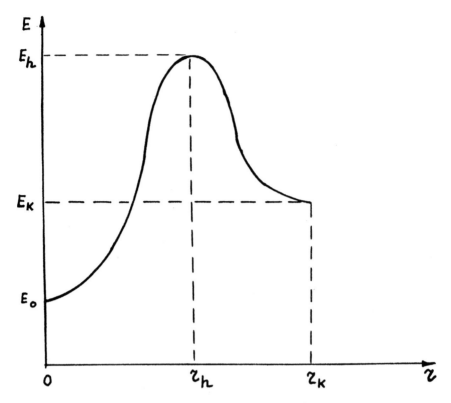

Figure 3. The dependence of the total energy of particle on the coordinate of reaction r.

particle concentration, determined by K_{eq} and initial concentrations of molecules (particles), forming intermediate particle, and the probability of the existence of such particles with energies, higher than E_n

$$\left[p = A \cdot \sum_{E_i = E_n}^{E_i = E_{max}} \exp(-E_i / RT) \right].$$

Motion time of intermediate particle, possessing E_n energy, calculates all the time, required for climbing, as well as for lowering of this particle from the reaction barrier. As these reactions proceed preferably in the gas phase, E must calculate the influence of solvent (or the surrounding of intermediate particles) on the motion process. The probability of particle lowering from the top of the barrier, determined in accordance with the fluctuation theory, equals:

$$P_k = \left\{ 1 + \exp\left[-\frac{dE/dr}{RT} \right] \right\}^{-1} .$$

In the cases, when the dependence $E=f(r)$ is not presented in clear view, it is supposed, that the motion of intermediate particle by the reaction barrier is equally accelerated (lowering) and equally decelerated.

Considering the present model, parameters of just intermediate particles enter the expression for the rate constant of the reaction, characteristics and structure of which can be estimated, basing on system thermodynamics, and the rate constant itself does not depend on the reaction order (the motion over the barrier of intermediate particle). The observed reaction order and rate constant dimension depends on equilibrium of forming intermediate particles, as in thermodynamics. It is necessary to point out, that in this case the rate constant depends on both initial and end state. Concurrent with the described theories about rate constants of chemical reactions there exists in literature [13] a number of others, basing on which elementary chemical act is considered more or less a collision and motion of particles over the reaction barrier.

All models of the rate constant lead to the expression, relative to Arrhenius equation

$$k = k_o e^{-E_a / RT} \tag{10}$$

where k_o - collection of parameters, depending weakly of the process temperature. However, it is probable that particular models can be used only for definite temperature ranges, that must lead to different physical sense of k_o and consequently, to the possibility of deviation from linear anamorphose of Arrhenius equation. Thus at low temperatures (equation 9) temperature dependence of the rate constant in defined in general by that of the equilibrium constant of the formation of intermediate particles and by probability of existence of such particles, possessing energies, enough for energetic barrier overcoming.

At moderate temperatures (the theory of activated complex) temperature dependence is generally defined by the influence of temperature on the change of motion rate of particles, colliding through activated complex. In consequence of difficulty of theoretical calculation of the reaction rate constant according to any model, the values of constants are defined, basing on the law of acting masses from experimental data under the supposition of ΔH, E_a, S constancy. Considering

wide temperature range in accordance with the dependence of thermodynamic parameters of the formation (H, S) of compounds on temperature, it is necessary to take into account temperature dependencies of
the present characteristics, also. The consideration of process thermodynamics and experimental data shows, that at least a number of
chemical reactions appears, starting from definite temperature. At the
same time the analysis of temperature dependencies of the rate constants of the reaction do not lead to such conclusion, i.e. to the conclusion about the reaction stop at definite temperature. Such contradiction
may be abolished by the introduction of the supposition about the fact,
that at least at the reaction initiation it is reversible. Further on at
temperature change one of them becomes to prevail. The restriction of
the values of the reaction rate constants in dependence on temperature
is caused by the occurrence of new processes, or the rate constant
reaches the value, defined by frequency of system oscillation (separate
parts of the molecule).

It is known [14-18] that the most number of degradation chemical
reactions proceeds through the number of elementary chemical acts. In
this case taking into account supposed reaction mechanism and the
principle of independence of the chemical reaction, the rate of the
change of each component (forming and destroying) is considered as the
sum of rates of proceeding elementary acts [19]. For the number of chemical reactions, for example, for conjugated ones, the principle of independence of chemical reactions is not fulfilled. It is not clear, that the
number of differential equations, characterizing the rates of component
change, relates to the number of components, existing during all the
process, i.e. the number of equations relates to the number of variables.
To simplify the solution, the part of these differential equations may
be substituted by the same amount of algebraic equations of the material balance. Solving the system of kinetic equations the time dependence of the components occurs; in this case as some components participate in several elementary acts, and the number of measuring parameters of the process is usually limited, it becomes possible to estimate
the combination of elementary act constants, and not themselves. It is
clear, that determining values of activation energies in these cases are
also seeming (observable), and not real ones. Different regularities of
chemical acts, observed in chemical processes, may be reduced to three
types of reactions: reversible, regular and parallel ones and their combinations. Principles of writing down of such nonelementary chemical
reactions are displayed in every manual on physical chemics. The
main features of polymer reactions, including degradation reactions, are
stipulated by multitude of reactionary centers (unitypical often), by
the change of mobility of the system (chain flexibility) and its form
(volume) [4,6]. All effects, connected with these features of the change
of chain molecule structure, may be displayed simultaneously, but in

some cases prevails one of them. The change of mobility and system form usually leads to the change of reactionary ability (rate constants) of active centers, rarely to the change of the amount (concentration) of active centers. These features of degradation processes proceeding will be considered below in corresponding papers. The multitude of reactionary centers of the chain stipulates the formation of compounds during reaction, which differ by disposition and the amount of particular chemical groups or size (molecular mass). That is why there appears the necessity to describe processes, connected with the formation or the destruction of compounds (chains) of definite size and structure. In this case it is supposed, that the regularity of elementary chemical acts of processes, proceeding in different places of the chain, is equal. Evidently, describing kinetics of processes the combination of both regarding forms must be used, i.e. there must be taken into account both chemical mechanism of the process and the place of the reaction proceeding.

Three types of molecules exist in the reaction:

1) initial molecules, which are destroyed during the process. The rate of destruction of these molecules is described by the totality of the reaction rates, proceeding by all active reactionary centers, under the supposition of the independence of chemical reactions.

2) Intermediate molecules, which can be formed from initial and intermediate molecules and be destroyed resulting the reaction on them. The rate of formation (destruction) of such compounds represents the sum of formation rates of the present molecules by all possible variants and the difference of the reaction rates, leading to their destruction.

3) Final reaction products, which cannot be destroyed during the process and are formed from large ones only. The rate of their formation is described by the totality of the reaction rates of initial and intermediate molecules, which lead to the formation of the particular reaction product.

Particular forms of writing of such reactions are presented in the paper [19]. Examples will be given in corresponding papers. Probability character of the proceeding of uniform reactions leads to the occurrence of the possibility to describe the process statistically and thermodynamically [20]. Evidently, the description of structure or chain length by kinetic as well as by statistic method must give uniform result, which stipulates additional possibility of experimental checking of the supposed reaction mechanism by molecular mass (molecular weight distribution) or by the structure of chain parts (their distribution).

The statement about the reaction mechanism is based on kinetic description of the supposed process and its coincidence with experimental data. The accuracy of experimental data and calculation methods may lead to coincidence with experiment not a single scheme. In this case there exist no criteria of perfect scheme selection. That is why it was supposed the principle of minimum of performing labor [10] for the introduction of additional demands to the process, i.e. the rate of free energy change during the process must be maximum. The principle is similar to the demand of minimum of the system formation free energy, which application is a must under the consideration of equilibrium systems. As it was mentioned in the paper [10], the application of the principle of performing labor maximum allows to estimate the most probable process mechanism from the suggested ones (in thermodynamics the minimum of free energy is determined at set system composition), and to estimate the correlation between different reaction mechanisms, which may proceed in the system.

Though any process must be described with regard to kinetic and thermodynamic parameters, at present time there is no strictly stated connection between kinetics and thermodynamics. To set such connection it was suggested to introduce the measure of chemical reaction inertness at kinetic consideration [21]. Then the motion force of the process is the change of free energy, and reaction acceleration may be characterized by rate constant and concentration of reacting substances. Unfortunately, we do not know any experimental works, which prove or refuse the suggested connection between thermodynamic and kinetic parameters of the process.

REFERENCES

1. E.F. Vainshtein, G.E. Zaikov, *International Journal of Polymeric Materials*, 1995, in press.
2. N.M. Emanuel, D.G. Knorre, *Chemical Kinetics* (in Russian), 4th Edition, Moscow, High School Publishers, 1984, 463 pp.
3. N.M. Emanuel, G.E. Zaikov, V.A. Kritsman, *Chemical Kinetics and Chain Reactions*. Historical Aspects, New York, Nova Science Publishers, 1995, 685 pp.
4. *Kinetics and mechanism of Formation and Transformation of Macromolecules* (in Russian), Ed. by N.A. Plate, Moscow, Nauka (Science) Publishers, p. 273.
5. *Chemical Reactions of Polymers* (in Russian), Ed. by V.A. Kargin, Vol. 1, Moscow, Mir (World) Publishers, 1967, 546 pp.
6. E.F. Vainshtein, DSc. Thesis, Moscow, Institute of Chemical Physics, 1981, 450 pp.
7. V.A. Afanasiev, G.E. Zaikov, *In the realm of Catalysis*, Moscow, Mir Publishers, English version, 1979, 120 pp.
8. K. Leidler, *Kinetics of Organic Reactions* (in Russian), Moscow, Mir Publishers, 1966, 348 pp.

9. S. Gleston, G. Eiring, K. Leidler, *Theory of Absolute Rate of Reactions* (in Russian), Moscow, Mir Publishers, 1968, 569 pp.
10. E.F. Vainshtein, *Mathematical Methods in Chemical Kinetics and Theory of Combustion* (in Russian), Kyzyl, Tuva Publishers, 1991, p. 27.
11. L.A. Blumenfeld, *Problems of Biological Physics* (in Russian), Moscow, Nauka (Science) Publishers, 1974, 446 pp.
12. E.F. Vainshtein, *Russian Journal of Physical Chemistry*, 1989, vol. 63, No. 9, p. 2838-2846.
13. H. Westerhoff, K. Van Dan, *Thermodynamics and Regulation of Transfer of Free Energy in Biological Systems* (in Russian), Moscow, Mir, 1992, 684 pp.
14. N.M. Emanuel, A.L. Buchachenko, *Chemical Physics of Polymer Degradation and Stabilization*, Utrecht, VNU Science Publishers, 1987, 340 pp.
15. G.P. Gladyshev, Yu.A. Ershov, O.A. Shustova, *Stabilization of Thermostable Polymers* (in Russian), Moscow, Khimia (Chemistry) Publishers, 1979, 312 pp.
16. V.Ya. Shlyapintokh, *Photochemical conversion and stabilization of polymers*, Munich, 1984, 420 pp.
17. Yu.A. Shlyapnikov, S.G. Kiryushkin, A.P. Mar'in, *Antioxidative Stabilization of Polymers*, London, Ellis Harwood, 1995, 382 pp.
18. V.A. Afanas'ev, G.E. Zaikov, *Physical Methods in Chemistry*, New York, Nova Science Publishers, 1992, 182 pp.
19. E.F. Vainshtein, Kinetic Aspects of Polymer Degradation, In: *Polymer Yearbook*, Ed. by R.A. Pethrick and G.E. Zaikov, London, Harwood Academic Publishers, 1995, vol. 12, p. 15-23.
20. N. Grassie, *Degradation of Polymers* (in Russian), Moscow, Inostrannaya Literatura (Foreign Literature) Publishers, 1969, 288 pp.
21. E.F. Vainshtein, *Oxidation Communications* (Bulgarian-English Journal), in press.

Thermodynamic Opportunity of Degradation Reaction Initiation

E.F. Vainshtein and G.E. Zaikov
Institute of Chemical Physics, Russian Academy of Sciences,
4 Kosygin Str., Moscow 117334, Russia

Motive power of any process, including chemical degradation, is the change of free energy ΔG. Reaction proceeding becomes possible if

$$\Delta G = G_e - G_{in} < 0 \tag{1}$$

where G_e, G_{in} - free energies of end and initial system state formation. Condition of thermodynamic opportunity of the reaction initiation, which performance is defined by kinetic parameters, is set by the equation

$$\Delta G = 0. \tag{2}$$

Practically always there occur the conditions, at which regularity (1) is fulfilled, at storage and exploitation of polymeric materials [1-9] in consequence of external parameters change. Lets consider the conditions, fulfilled parameter (2) for polymers, existing in different phases, at which degradation process proceeding can be initiated in the system. We can restrict by consideration of thermodynamics (formation free energy) of the initial and end state, in which definite chemical substances are contained, because the end state at the defined initial one does not depend on the way of its reaching. I.e. it is possible not to take into account the detailed mechanism of the process to clear up thermodynamic opportunity of its proceeding. Let consider proceeding probability of the following reactions:

a)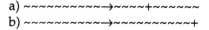
b) ~~~~~~~~~~~~→~~~~~~~~~~~~+

c) ~~~~~~~~~~~~+O_2→~~~~~~~~~~~~~→~~~~~OR+R_1O~~~~~
d) ~~~~~~~~~~~~+O_2→~~~~~~~~~~~~→~~~~~~~~~~OR+R_1O~~~~

Reactions (a) and (c) describe the process of chemical bond break in the main chain, and (b) and (d) - in side chain. Reactions (a) and (b) are characteristic for thermodegradation, and (c) and (d) - for thermooxidation degradation. It is evident, that reactions (a) and (b) differ sufficiently from (c) and (d) by potential mechanism of proceeding. In particular polymers under thermodynamic consideration it is necessary to take into account the process of the first attachment of oxygen (or other chemical agents) only. Further on it will proceed chemical transformation of groups, neighbor to attached oxygen, followed by chain fracture. That is why we consider, first of all, conditions of thermodynamic opportunity of the initiation of (a) and (b) type reactions. In dependence on the place of its proceeding first reaction will occur with a sufficient decrease of molecular weight at breaking by chain middle. If the chain break occurs by the end, the reaction will be similar to that of (b) type. We will consider combining free energies additive, that in some particular cases can lead to different parameters, at which the process occurs. Probability of such consideration is connected with independence of free energy of end and defined initial system state formation. I.e. in this case as though it is composed the way of transition for each stage, during which only one component of free energy changes. First of all, let consider the simplest case - thermodynamic opportunity of degradation reaction initiation in diluted solution.

1.1. LIMIT SIZES OF THERMODYNAMICALLY STABLE CHAINS IN DILUTED SOLUTION

Free energy of solution, containing chain molecules of a definite length (before the beginning of degradation), can be written as follows:

$$G_p=G_x+G_s+G_{id}+G_{ex}+G_{defl}+\Delta G_s+\Delta G_{p-s} \qquad (3)$$

Here G_x is free energy of chain molecule formation, representing "rigid bar"; G_s - free energy of solvent formation; G_{id} - free energy of ideal mixing; G_{ex} - excessive free energy of different size molecules mixing (polymer and solvent), G_{defl} - free energy of chain deflection, ΔG_s - change of free energy of solvent molecule interaction with each other at transition from solvent volume to the chain (Figure 1). This interaction is stipulated by the change of interatomic distance in solvent molecules, disposed in polymer volume among the chain. ΔG_{p-s} - free energy of chain-solvent interaction (precisely, the change of free energy in consequence of interaction between solvent molecules in bulk and chain molecules in mass (by local interactions of solvent molecules with the chain).

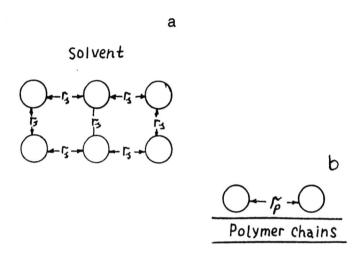

Figure 1. The distance between solvent molecules in volume (r_s) - a. The distance between solvent molecules disposed neighbor to the polymer chain (r_p) - b.

Free energy of solution formation, consisted with the same amount of solvent and degradated molecules, is presented as follows:

$$G_e = G_x + G_s + G'_{id} + G'_{ex} + G'_{defl} + \Delta G'_s + \Delta G'_{p-s} + \Delta G_x \qquad (4)$$

Indexes in this equation mark the same components, as in (3), but relate to a new state.

$$\Delta G_x = \Delta H - T\Delta S$$

- the change of free energy at new bond formation. We can neglect the contribution of bulk into free energy change, owing to its low change at degradation in diluted solution. We obtain then:

$$\Delta G = G_e - G_p = (G'_{id} - G_{id}) + (G'_{ex} - G_{ex}) + (G'_{defl} - G_{defl}) +$$
$$+(\Delta G'_s - \Delta G_s) + (\Delta G'_{p-s} - \Delta G_{p-s}) + \Delta G_x \tag{5}$$

At similar reaction with the participation of low molecular compounds only ΔG will be the combination of all components, except $(G'_{ex} - G_{ex}) + (G'_{defl} - G_{defl})$, values of $(\Delta G'_s - \Delta G_s) + (\Delta G'_{p-s} - \Delta G_{p-s})$ differing from the ones of the same components in equation (5) among the chain. In the most number of cases in low- and high-molecular reactions these terms possess similar values, because they characterize local interactions, i.e. that between the closest atoms. Designing these terms as G_{pp}, we obtain

$$\Delta G = \Delta G_{pp} + (G'_{ex} - G'_{ex}) + (G'_{defl} - G_{defl}). \tag{6}$$

Consequently, the features of degradation proceeding in diluted solutions of chain molecules in comparison with similar reactions in low molecular compositions will be stipulated by the change of chain flexibility and number of polymer-solvent contacts. The change of both components is stipulated by the change of chain length (by its shortening). The present consideration does not take into account the existence of rotary isomers in chains and their changes in consequence of chain length change. If consider them, free energy of both initial and end state should be added by terms, taking into account their amount and probability of various disposition in the chain. At the change of correlation between conformers the changes of flexibility and contact number should be taken into account in the process.

For infinitely long chains (polymers) terms $(G'_{ex} - G_{ex}) + (G'_{defl} - G_{defl})$ will be close to zero. Consequently, in infinitely long chains and polymers free energy change in the process can be close to the one in similar reactions of low molecular compounds (P. Flori principle) [10].

If ΔG_x values are sufficiently lower (greater by absolute value) than the rest of components of free energy change, the process can proceed at all the value of molecular weights from the point of view of thermodynamics. Consequently, their application in the definite conditions is unadvisable.

$(G'_{id} - G_{id}) < 0$ for selected mechanism, i.e. in the whole range of molecular weights. The change of free energy of ideal mixing is favorable to degradation process proceeding.

$$G'_{id} - G_{id} = -RT\left[\frac{1}{2X'_2 X_1}\ln\left(\frac{X_1}{X_2}\right) + \ln(1 + X_2) - X'_2 \ln 2\right] \tag{7}$$

Here X_1, X_2, X'_1, X'_2 - molar parts of solvent and chain molecules in initial and end (') states. Let us remind, that in calculation per mole of double mixture

$$G_{id} = -RT\sum_i x_i \ln x_i,$$

where x_i - molar parts of components. It is necessary to introduce corresponding coefficient into the equation, if changes of concentrations and number of moles are taken into account. The value of $(G'_{id} - G_{id})$ is constant at constant initial concentration and number of breaks of chain molecules. Values of $(G'_{id} - G_{id})$ for a definite number of chain breaks (more than one) can be estimated in analog. G_{p-s}, characterizing free energy of polymer-solvent interaction for sufficiently long chains, can be expressed as follows:

$$G_{p-s} = C[n_c q_{p-s}^{c+}\sum_i t_i q] \tag{8}$$

where C - chain concentration; q_{p-s}^c - molar free energy of local interaction of middle chain units, n_c - their number for end groups, also (if the chain is chemically inhomogeneous, summation of free energies of all monomeric units is required). Chain inhomogeneity and solvent contacting with it may be stipulated by chemical defects as well as by rotary isometry. For insufficiently long homogeneous chains G_{p-s} becomes a linear function of molecule number; q_{ie} - local free energy of interaction for different end groups t. If there are two ends, and they are chemically equal, then

$$G_{p-s} = C[n_c q_{p-s}^c + 2t q_{p-s}^G]$$

At small chain lengths G_{p-s} depends nonlinearly on the chain length (the number of monomeric units), as a consequence of its deflection (chain form does not represent "statistic tangle" or "rigid bar"). In presence of conformers it occurs in consequence of conformatic composition dependence on chain length. This component can be written down similarly after degradation. In this case it is necessary to taken into account, that the number of molecules becomes twice higher, but total length of both chains, in the ranges of undertaken suppositions, differs from the initial degradating chain by one monomeric unit, and the number of ends becomes twice greater for each chain. For example, at degradation of complexes of AlR_3 polyethylene glycole end groups ~$OAlR_2$ and $R\text{-}CH_3$ are formed in addition to two initial ones. They are formed according to the equation:

$$HO \sim\sim\sim\sim CH_2 - O - CH_2 \sim\sim\sim\sim OH \rightarrow CH_2OAlR_2 + RCH_2 \sim OH$$
$$AlR_3$$

It is clear that this value relating to a single break does not depend on concentration and chain length and is similar to low molecular compounds, beginning from a definite length of the initial chain only. In this case it was supposed, that molecules of solvent and chains do not interact with each other in the bulk. In more general case it is necessary to take into account the fact that at solving $G_{p\text{-}s}$ is expressed by the following equation:

$$G_{p\text{-}s} = nG_{s\text{-}s} + mG_{p\text{-}p} - lG_{p\text{-}s},$$

where $G_{s\text{-}s}$, $G_{p\text{-}p}$, $G_{p\text{-}s}$ - free energies of interaction; n, m, l - numbers of contacts between them. $G_{p\text{-}s}$ is the difference of free energies of contact formation, because chains represent diluted solution before and after degradation in accordance to accepted suppositions, i.e. they do not interact with each other. Consequently, $G'_{p\text{-}s}$, $G_{p\text{-}s}$ should include the change of free energy, stipulated by the interaction of solvent molecules with each other, which become free resulting the process.

For relatively long chains average value of interaction energies of solvent molecules, surrounding the chain, is independent on the chain length. One more additional component of free energy of chain formation occurs, because the distance between solvent molecules is defined by both interaction with the chain and their disposition among the chain. Figure 1 shows directions of interaction of solvent molecules, disposed neighbor to the chain, with the chain and with each other. Similar interactions in processes with the participation of low molecular components only can differ by their value. Basing on the above mentioned facts, it may be decided, that $\Delta G_{in\text{-}in}$ term can differ insufficiently from similar value of the process, proceeding with the polymer participation. As equilibrium constants, determined from the change of Gibbs free energy during the process, are related to a mole of reacting substance, then the components of free energy change will be considered further on, taking account of this fact. Free energy of homogeneous isolated chain deflection depends on its local rigidity and length, and may be characterized by the correlation ($\overline{R}_o^2 / \overline{R}_1^2$) of mean-squares of the chain distances at "free rotation" of bonds and the chain, being studied [11]. The dependence of deflection entropy of isolated chain, related to monomeric unit are shown on Figure 2, and the dependence of $\overline{R}_o^2 / \overline{R}^2$ on chain length - on Figure 3. It is seen from Figure 2, that accompanying the growth of isolated chain length this component, related to a mole, approaches the limit (R), stipulated by "free rotation" of bonds (by their statistic independence). The higher local rigidity is, at larger lengths (molecular weights) this limit is reached. $S_{ch}=0$ for very small chain lengths of simple structures, and $S_{ch/p}=R$ for very large ones. Local rigidity decreases at temperature increase. Consequently, the higher temperature is at shorter chain length the limit is reached.

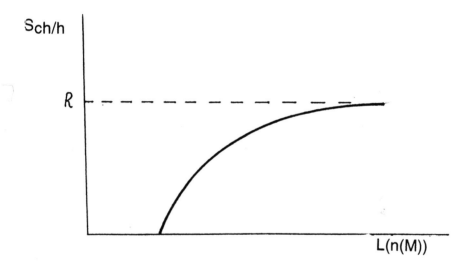

Figure 2. The dependence of deflection entropy of isolated chain related to monomeric unit (bond) S_{ch}/n on chain length L (n,M).

Solvent disturbs chain deflection, interacting with it: at first, solvent molecules must be moved at chain deflection. In this case is may decrease sufficiently the distance between interacting atoms, which are closely disposed and valently disconnected. The model of homogeneous chain in solution is described in [12]. The chain in the solvent possesses

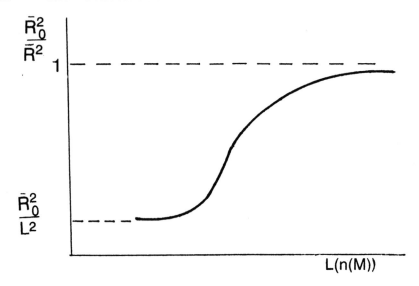

Figure 3. The dependence of $\overline{R}_0^2 / \overline{R}^2$ on chain length L.

higher observable local rigidity. As observable rigidity it is meant local rigidity of the isolated chain, possessing contour length and mean-square of the distance between chain ends the same as in the solvent. Paper [13] showed the following value of observable rigidity of homogeneous chain a_n:

$$a_n = \left(40a^2\right)/\left(40a - L^2 p\right). \tag{9}$$

Here a - local rigidity of isolated chain, determined on the basis of potential hole, L - its contour length, p - interaction parameter, characterizing the influence of the solvent on chain deflection. For more details about parameter p and its physical sense see [12]. As it is seen from the equation (9), the greater chain length is, the higher a_n value is at the definite temperatures. It is evident, that at a definite value of chain length (L_c) denominator becomes zero, and a_n becomes infinite. It means physically, that the chain cannot exist at the length greater this value ($a_n < 0$). Consequently, any solvent must show the chain length, larger which value the chain is instable thermodynamically. The value of L_c equals

$$L_c = \sqrt{40a/p}. \tag{10}$$

It is seen from the equation (10), that the larger p is and smaller a is, the lower critical chain length is. The dependence of L_c and a and p is shown on the Figures 4 and 5. However, it should be considered the value of free energy of molecule deflection in solution, because length of thermodynamically instable chain is defined by total change of free energy. If the chain represents "rigid bar", as it was mentioned above, entropy of its deflection equals zero. If there exists "free rotation" of bonds in the chain, free energy of deflection, related to monomeric unit (bond), equals RT. As entropy is characterized by the number of chain states, its value in the first approximation can be characterized by correlation of mean-squares of distance between chain ends at "free rotation" of the chain in the solvent (precisely the correlation of distribution functions). If $\overline{R}_0^2$ can be calculated easily for the chains of any structure, $\overline{R}_s^2$ may be determined experimentally, for example, by viscosity. As the view of S_{ch}/n function in general case is unknown, it may be written down as the expansion:

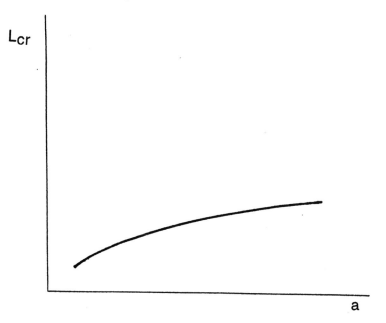

Figure 4. The dependence of limited length of thermodynamically stable chains L_{cr} on its local rigidity a.

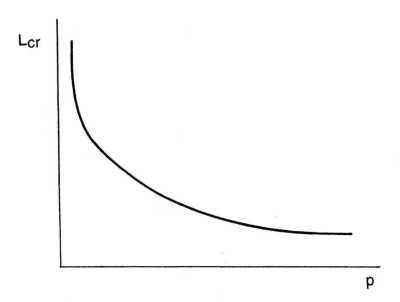

Figure 5. The dependence of limited length of thermodynamically stable chain L_{cr} on interaction parameter polymer-solvent p.

$$S_{ch}/n = a + b\frac{\overline{R}_0^2}{\overline{R}^2} + c\left(\frac{\overline{R}_0^2}{\overline{R}^2}\right) + \ldots$$

Usually for noncyclic chains $\dfrac{\overline{R}_0^2}{\overline{R}^2} \leq 1$.

Apparently, in the most number of cases the consideration may be limited by first three terms of the expansion with the accuracy, usually satisfying the experimental data:

$$S_{ch}/n = a + b\frac{\overline{R}_0^2}{\overline{R}^2} + c\left(\frac{\overline{R}_0^2}{\overline{R}^2}\right)^2.$$

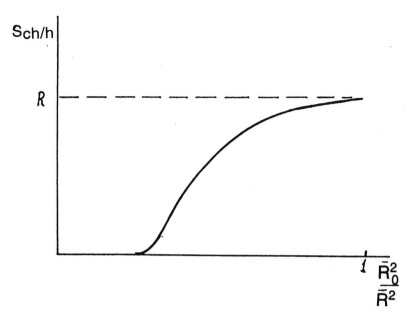

Figure 6. The dependence of deflection entropy related to monomeric unit, Sd_{efl}/n on $\overline{R}_0^2/\overline{R}^2$.

It is evident that this function is continuous one (see Figure 6). Consequently, it may be used the following additional correlations for the estimation of expansion coefficients for noncyclic chains:

$$0 = a + b\left(\frac{\overline{R}_0^2}{L^2}\right) + c\left(\frac{\overline{R}_0^2}{L^2}\right)^2$$

$$R = a + b + c.$$

For infinitely great chains a=0 and b+c=R. Coefficients b and c can be separated, $\overline{R}^2$ value being determined in any experimental point. One can use the expression, containing greater number of terms in the expansion, if required. Then the number of experimental determinations of coefficients will grow, consequently. For the chain (homogeneous), consisted with one conformer, osmotic forces are similar to the ones stretching the molecule by the ends. Then labor of the distance being estimated at constant temperature, one can estimate S_{ch} value from the condition $A_d = F \cdot \Delta\overline{R}^2 = T_n(\Delta S_{ch})$. If more accurate estimation of the distance change value between the ends is required, one may calculate

$\overline{R}$ value, taking into account distribution function of the distances between the ends $\overline{R}^2$. The value of the mean-square distances between the ends of the chain, being stretched by forces applied to the ends equals [14]:

$$\overline{R}_d^2 = 2 \cdot \int_0^L (L-l) \exp\left[-\left(RT/\sqrt{f_1 a_1} \cdot th(1/2 \cdot \sqrt{f_1/a_1})l) - RT/\sqrt{a_2 f_2} \cdot th(1/2 \cdot \sqrt{(f_2 l/a_2})\right)\right]dl$$

where l - particular coordinate; a_1, a_2 - values of coefficient a in different directions; f_1, f_2 - values of stretching force in different directions. The value of S_{ch}, characterizing the number of chain states, may be estimated also, knowing the distribution function of distances between the ends. Evidently, two new chains occur at the considered fracture:

$$L = 2L_1 + \Delta l,$$

where L_1 - single chain length; $L_1 + \Delta l$ - another chain length. If the initial chain is long enough, then $2L_1 \gg \Delta l$. Then it is possible to estimate the value of observable local rigidity of fractured parts and to compare the decrease of local rigidity:

$$a_o - a_{of} = \frac{120 a^2 L^2 p}{(40a - L^2 p)(160a - L^2 p)} = a_o \left[\frac{3L^2 p}{160a - L^2 p}\right],$$

where $40a > L^2$.

The equation describes smoothly increasing function, i.e. the value of differences of observable local rigidities increases with the chain length sufficiently faster, than a_o of the initial chain (Figure 7) (increasing coefficient $3L^2 p/[40a-L^2 p]$). It should be mentioned that entropy of isolated chain deflection must increase according to the applied model with the increase of the chain length and approaches RT. The value of local rigidity of initial and broken isolated chains are equal, and entropy change in the solution according to initial chain length is extreme function (Figure 8).

In fact, at the consideration of this formula [11] is supposed that both new chains are close by sizes. if obtained chains differ sufficiently by sizes, their observable local rigidities are advisable to be estimated by the formula (9). The dependence of the chain entropy on

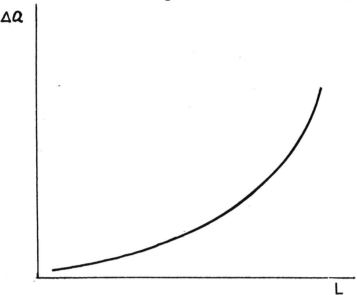

Figure 7. The dependence of changing of observable rigidities (Δa) of rupted and initial chain on initial chain length L.

its length is shown on Figure 2, according to Landau-Livshitz model [12] and observable local rigidities on p and L - on Figures 9 and 10. If $\overline{R}_o^2$ is reached for isolated chain at infinite length, then such length cannot be reached in the solvent, as it was mentioned above.

$$\frac{S_{ch/s}}{2n}(2n-1) - \frac{S_{ch/p}}{n} \cdot n = \left(\frac{2S_{ch/s}}{2n} - \frac{2S_{ch}}{n}\right) \cdot n - \frac{S_{ch/s}}{2n} > 0$$

For relatively short chains

$$G'_{defl} - G_{defl} = -T\left(n\left[\frac{2S_{ch/s}}{2n} - \frac{S_{ch}}{n}\right] - \frac{S_{ch/s}}{2n}\right)$$

for infinitely long chains n>>1

$$G'_{defl} - G_{defl} = -Tn\left(\frac{2S_{ch/s}}{2n} - \frac{S_{ch}}{n}\right) \tag{12}$$

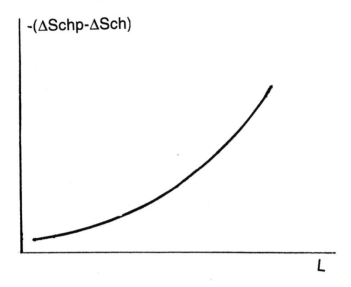

Figure 8. The dependence of entropy change (ΔS_{chp}-ΔS_{ch}) on initial chain length L.

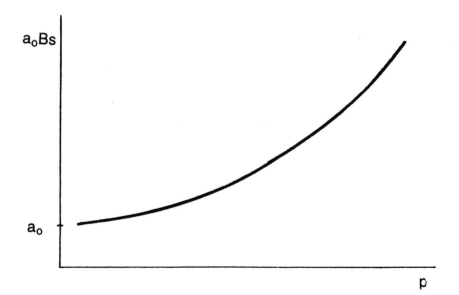

Figure 9. The dependence of observable local rigidities of chain (a_{obs}) interaction parameter polymer-solvent p.

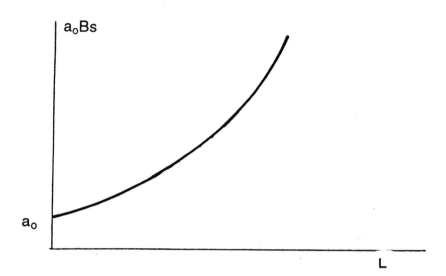

Figure 10. The dependence of observable rigidity of chain (a_{obs}) on chain length L.

As (a_0-a_{of}) increases with the chain length, (G'_{defl}-G_{defl}) increases with higher rate (with higher derivative):

$$d(G'_{defl}-G)/dL.$$

Let us use Flori-Haggins formula [14,15] to estimate the change of mixture excessive entropy:

$$G_{ex} = -RT\left[x_1 \ln \frac{\phi_1}{x_1} + x_2 \ln \frac{\phi_2}{x_2}\right],$$

where $\phi_i = r_i x_i / \sum_i r_i x_i$ - volumeric part of components i=1,2,r2=r - the number of places, occupied by chain molecule in condition, that solvent occupies one place in it ($r_1/r_2 \sim V_1^o/V_2^o$). The result, characterizing physical sense of the process, is obtained also at the consideration of Guggengame-Stavermann theory [16-18] and more accurate expressions of Prigojin's theory [19]. As we are interested just in thermodynamic opportunity of degradation reaction, the model, simplest and generally accepted for polymers, was selected for the estimation of excessive entropy of different size molecules mixing. It is necessary to point out that Prigojin's model, applying contact number, is suitable for the chains of limited size molecules mixing, which are contained before the process initiation, equals:

$$G_{ex} = -RT\left\{\frac{N_1}{N_1+N_2}\cdot\ln\left[\frac{N_1+N_2}{N_1+rN_2}\right]+\frac{N_2}{N_1+N_2}\cdot\ln\left[\frac{r(N_1+N_2)}{N_1+rN_2}\right]\right\}.$$

After degradation the number of molecules becomes twice higher, and if it is supposed, that the break occurred at the middle of the chain, approximately, then it can be written:

insert equation here

It was supposed at the deduction, that the number of places in quazicrystal cell decreases twice. Such supposition describes perfectly relatively large chains, where $(r/2+t)\approx r/2$. Here t - the difference in the number of solvent molecules, surrounding the chain. In comparison with $r/2$, t is stipulated by contacts of the solvent with the ends, forming at chain breaks. Then

$$G'_{ex} - G_{ex} = -RT[\ln(1+x_2)+(1/2\cdot x_1x_2'\ln r) - x_2\ln r] \qquad (13)$$

It is seen from the equation (13), that $(G'_{ex} - G_{ex})$ should increase by absolute value according to chain length (r increases) at constant concentration of the initial chains. As terms of free energy change, characterizing variation of chain nature of a molecule, decrease according to molecular weight, then independently on terms, characterizing the change of local interactions, molecular weight will be reached, at which $\Delta G=0$ and, kinetic opportunities existing, degradation proceeding will start. But if there is no possibility for the degradation proceeding according to kinetic causes, then according to thermodynamics the growth of molecules must be limited. The value of molecular weight, at which the break occurs, is defined by total selection of the components of the process free energy change, i.e. by the change of bond energies, temperature, solvents. The weater the bond change is, the higher molecular weight provides the occurrence of the chain degradation. Lets consider the present dependence on the example of degradation of simple ether polymeric complexes with AlR_3. Thus according to calorimetric data, complexes of $Al(CH_3)_3$ with polyethylene glycole starts from $\overline{M}_n = 5000$, complexes of $Al(C_2H_5)$ - from $\overline{M}_n \sim 6000$. For complexes of $Al(C_2H_5)_3$ with polyformaldehyde the degradation is observed at $\overline{M}_n \sim 800$, for complexes of $Al(CH_3)_3$ - at $\overline{M}_n = 525$ [13]. The reaction was performed in benzene solution with polymers, possessing sharp molecular weight distribution. At present is may be taken established, that for the chains of any structure, particularly, there exist extreme molecular weights of the chains in diluted solution. On the whole, these extreme molecular weights are the consequence of degradation processes, however the limiting of their growth is probable. In

this way is was shown [20] the occurrence of $-\overset{\displaystyle H \quad\ H}{\underset{\displaystyle H \quad\ H}{C-C}}-$ bond degrada-

tion. Growth limiting is usually characteristic for extremely rigid chains with large bond energies, for example, for the ones with the conjugation system [21]. Solvent influence and the value of the chain length, at which thermodynamic opportunity of degradation process proceeding occurs, is comfortable to be considered on the example of influence on separate components of free energy change, as well as on their combinations. Evidently, the picture of the occurrence is very diversified, because the solvent influences practically all components (except $G'_{id}-G_{id}$). The influence on ΔG_x is stipulated by the change of chain surrounding by solvent at new end groups formation. Solvent molecule size influences most sufficiently $(G'_{ex}-G_{ex})$, because r decreases with the solvent molecule size growth. Consequently, the larger solvent molecules are, at greater chain lengths thermodynamic opportunity of polymer degradation occurs, in other similar conditions. Local interaction of the solvent with the chain is expressed by the change of depth, as well as by the form of potential pit $(\Delta G'_{p-s}-\Delta G_{p-s})$ (the change of observable local rigidity) and by the change of the interaction of solvent molecules with each other, surrounding the chain $(\Delta G'_s-\Delta G_s)$. As $(\Delta G'_{p-s}-\Delta G_{p-s})$ and $(\Delta G'_s-\Delta G_s)$ are connected linearly with the chain length (in the first approximation for sufficiently long chains), and the change of observable local rigidity leads to linear dependence of the change of this free energy component on the chain length, then their break is probable according to the chain length, evidently. Moreover, local interactions (their components) are usually stipulated preferably by energetic components. The change of deflection free energy is stipulated preferably by entropy change.

Temperature influence is reflected in all components of free energy change. The changes, related to ΔG_x, $(G'_{id}-G_{id})$, are similar to the ones, observed in the processes with low molecular compounds participation, only. They are described in detail in many manuals on physical chemistry, thermodynamics and kinetics. At temperature increase the

changes among molecules, surrounding the chain, are probable in consequence of local interaction weakening. This may lead to the change of
their interaction and to the decrease (increase) of the number of chain-
solvent contacts, and consequently to the change of ($\Delta G'_s$-ΔG_s) and
($\Delta G'_{p-s}$-ΔG_{p-s}) components. If we accept the changes of the rest of components to be insufficient at temperature change, that is wrong in many
cases, even in these conditions there is observed no linear dependence of
ΔG on temperature. Under the extreme conditions energies of local interactions will approach zero (solvent molecules will be separated by
sufficiently long distance from each other). If the interaction intensity
equals zero, the system represents a rarefied gas. In this case association is not taken into account. Usually these deviations are small, but
the ones, related to (G'_{ex}-G_{ex}) change may become determinant in ΔG
change. This term changes in two ways by means of T coefficient in the
equation (13), as well as by means of r change. It is necessary to men-
tion, that if T changes in one direction, increases, for example, r may
decrease. Usually r decreases smoothly according to T growth, approaching constant value. At T>0 (the case, which is not fulfilled in
practice) the value of r is defined by the length of stretched chain and
by sums of Van-der-Vaals radii of contacting atoms. At T→∞ (the case,
which is not practically fulfilled also) r value is defined by minimally probable density of solvent, in the supposition, that distribution
of solvent molecules in the volume is equally probable. If it is supposed, that the rest of parameters are constant, extrema may exist in
the system, although they occur very rare, apparently. It is most probable, that in the most number of cases thermodynamic opportunity of
degradation occurs at temperatures, below which extremum is observed.
Extremum values of ($\Delta G'_{ex}$-ΔG_{ex}) can be observed in dependence on
temperature at the correlation between T and r, expressed by the equation:

$$r = e^{(c/(T-\Delta))}$$

where c - integrity constant, determined from experimental data;
$\Delta = ln[(1+x_2)-x_2 ln_2(/1/2 \cdot x_1 x_2$. Thus, ($\Delta G'_{ex}$-$\Delta G_{ex}$) decreases abruptly
with the temperature (increases by its absolute value), i.e. at tempera-
ture increase the contribution of this term into the change of free energy of the process increases. Temperature change influences (G'_{ex}-G_{ex})
doubly, also. On one hand, at temperature increase local rigidity increases from infinity at T→0 ("rigid bar") to the value, defined by
"free bond rotation" in the chain. After reaching the temperature, at
which "free rotation" of chain bonds is observed, local rigidity becomes
constant. On the other hand, chain-solvent interaction is weakening,
and the mobility of solvent molecules increases. This leads to even
sharper increase of the observable local rigidity.

If it is supposed that chain molecules are not associated with each other in the definite concentration range, then $(G'_{ex}-G_{ex})$ is a function of concentration according to the dependence. This term of free energy change decreases (increases by its absolute value) with the concentration increase (equation 13). Consequently, at a definite ratio of free energy change terms thermodynamic opportunity of degradation reaction initiation may occur, beginning at a definite concentration. I.e. if there exist kinetic opportunities for reaction performing, the proceeding of chemical degradation reaction may be observed at definite polymer concentrations in solution, then. Evidently, as it follows from the equation (13) thermodynamic opportunity of initiation and display of the reaction at definite concentration depends on temperature and solvent molecule sizes. Considering the opportunity of similar reaction proceeding by chain end or side group (Scheme 1 (b)), we obtain the largest change, contributed by $(G'_{defl}-G_{defl})$ term into motive force (equation 5).

As the number of chain molecules is the same before and after reaction, then $(G'_{ex}-G_{ex})$ term does not change practically. Consequently, in this case there are no specific features of the process display, different from similar reactions of low molecular compounds and stipulated by their concentration. As $(G'_{defl}-G_{defl})$ depends on chain length and on the place of reaction proceeding, then thermodynamic probabilities of its initiation by end and middle of the chain will be different. The reaction, proceeding by the end, depends weakly on molecular weight. That is why the probability of its initiation will differ insufficiently from that for similar reactions with participation of low molecular compounds, only. For reactions, proceeding by side groups, the decrease (increase by absolute value) of $(G'_{defl}-G_{defl})$ term will occur in the case of formation of a sphere, possessing lower observable local rigidity, during the reaction. As this dependence on chain length is of extremal character, initiation and proceeding of the reaction at definite molecular weight is probable then. Border values of molecular weights can be estimated, basing on the expression $\Delta G=0$. The place of the reaction proceeding will be of a special meaning for such processes. Forming more rigid parts, the most probable place of the reaction proceeding will situate closer to the chain end, in more flexible ones - closer to the middle (Figure 11). That is why in some cases chemical reaction is probable not for all side groups. This fact should be taken into account, calculating thermodynamic and kinetic parameters of the process. If more rigid parts are formed during the reaction, then the chain itself will be more stable thermodynamically, and the reaction will be decelerated according to the chain length growth (in some cases it will stop fully)[I]. At the formation of more flexible parts in the reaction[I]. At the opposite influence on the change of local rigidity and it is necessary to take into account the probability of reaction proceeding on the chain itself according to the mentioned overall, i.e. it is necessary to control the opportunity of the reaction initiation at the chain itself, also. Estimating thermodynamic opportunity of the reaction initiation, proceeding in accordance with Scheme B, let us take into account the main demand of thermodynamics - about the independence of free energy change on the way of transfer from particular initial state to the end one.

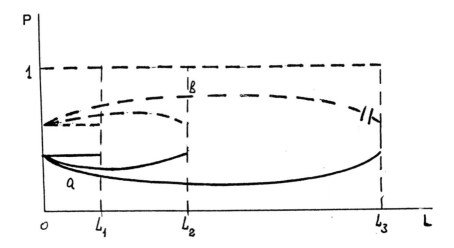

Figure 11. The dependence of probability of reaction proceeding on the active center position for different chain lengths: L_1 - low molecular chains, L_2 - oligomers, L_3 - polymer, a - the place after the reaction proceeding has higher observable local rigidity than initial chain. b - the place after the reactions proceeding has lower observable local rigidity than initial chain.

According to all abovementioned

$$\Delta G = \Delta G_1 + \Delta G_2,$$

where ΔG_1 - the change of free energy of transfer from initial state to intermediate one, ΔG_2 - free energy change at the transfer from intermediate state to end one.

Parameter ΔG_2 is similar to ΔG in reactions (a) and (b), and consequently is described fully by dependences, presented above:

$$\Delta G_1 = \Delta G_x + (G'_{ex} - G_{ex}) + (G'_{defl} - G_{defl}). \tag{14}$$

As the number of chain molecules does not change in this reaction, then $(G'_{ex} - G_{ex})$ is small in analog to (b) type reactions, and in many cases it may not be taken into account. At the formation of more flexible parts the reaction will proceed preferably by the chain middle, as in (b) scheme, and at more rigid one - by chain ends, preferably. If the reaction proceeds through the formation of intermediate complex, then it will occur most probably at the chain middle during the formation of more flexible complexes, and at the chain ends during the formation of more rigid intermediate chain-solvent interaction there may exist two probable places of the reaction, symmetric relative to the chain middle (Figures 12-14).

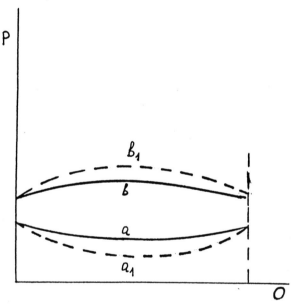

Figure 12. The dependence of probability of reaction proceeding on the active center position in the chain. a - the place after the reaction proceeding has higher observable local rigidity than initial chain. a_1 - the chain itself after reaction proceeding has higher local rigidity than initial chain during the formation of more rigid intermediate chain-solvent interaction. b - the place after the reaction proceeding has lower observable local rigidity than initial chain. b_1 - the chain itself after reaction proceeding has lower observable local rigidity than initial chain during the formation of more weak intermediate chain solvent interaction.

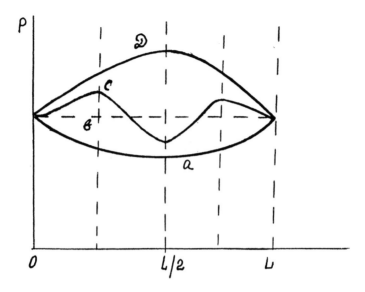

Figure 13. The dependence of probability of reaction proceeding on the position of active center in chain on different chain-solvent influence. The active center after reaction proceeding has lower local rigidity than initial chain: a - influence of solvent on deflection of reacted and unreacted place of chain the same. b - influence of solvent on deflection is compensated by influence of local rigidity. c - influences of solvent and local rigidities on deflection and equal, but have opposite direction of action. d - influence of solvent on deflection is stronger than influence of local rigidity.

Let remind, that chain fracture at small number of breaks is mostly probable at the chain middle. It is precisely the formation of intermediate reaction products and/or chain degradation, that stipulates reaction proceeding in the solution by the chain middle. Consequently, the feature of degradation reaction, as well as other chain ones, is the necessity of the reaction ability control of different reaction spheres. Evidently, the solvent can influence the correlation and disposition of conformers, that may change the most probable place of the reaction (Figures 13, 14). As the change of the most probable place of the reaction causes the obtaining of other products, it is necessary to take into account the probability of reaction mechanism change in polymer reactions by means of the change of chemical acts order (in analog to low molecular compound reactions), as well as the change of the reaction place at the same order of elementary chemical acts. Overall mentioned is related to thermooxidation degradation as well as to any other chemical reactions, proceeding through the formation of intermediate products. Thus at complex formation of AlR_3 with polyethylene glycole, at complex composition of (1:1) per monomeric unit further reaction of chain break proceeds preferably by the chain middle (complex formation and degradation coincides). At the proceeding of the reaction

$$\sim O - O \sim\to\sim\sim O - AlR_2 + R \sim\sim O \sim\sim$$
$$AlR_3$$

the break proceeds preferably by the end. Evidently the difference is observed for the chains of limited size. For polymers (infinitely long chains) break and complex formation becomes equally probable, i.e. the reaction proceeds according to chance law [13].

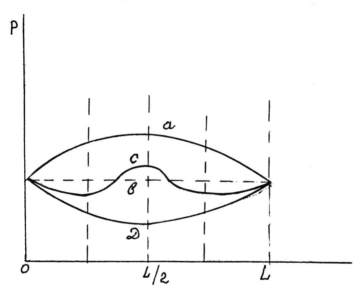

Figure 14. The dependence of probability of reaction proceeding on active center position in chain. Reacted place has higher local rigidity than initial chain: a - influence of reacted and unreacted places of chain on deflection the same. b - influences of solvent and local rigidity on deflection are equal but have opposite direction of action. c - influence of solvent on deflection is completely compensated by local rigidity influence. d - influence of solvent on deflection is predominate comparatively with local rigidity influence.

Association of macrochain molecules is displayed strongly as usual as the association of similar low molecular compounds. Free energies of deflection and mixing (of ideal solution and excessive one) are lost during association, but free energy of intermolecular bond formation increases. It is clear, that the longer the chain is, the higher probability of bond formation is. That is why, if one considers the initial state as the solution of associated initial molecules, and the end one - as the solution of fractured molecules, the change of free energy of the process will be lower then, than at isolated molecule break in the solution, because it is necessary to spend required free energy for associate break in order to obtain initial isolated molecules.

Fracture of initial chains, existing in association, is less probable thermodynamically even in the case, when the end state represents associate, including fragments, formed during initial molecule break. This is stipulated by the circumstance, that free energy of associates increases with the chain length growth. For example, if initial associate consisted with two molecules, and the associate in the end state contained three molecules (one of the chains broke into two parts), the associate is less stable in the end state, than in the initial one, because it can contain smaller amount of intermolecular bonds. Also at associate fracture in the end state great number of molecules is formed, and deflection entropy in the end state at any rate is not higher, than in the initial one (excessive and free energy of different size molecules mixing).

Evidently, in intermediate case of simultaneous existence of isolated and associated parts of fractured molecule in the end state the break will be less thermodynamically profitable, also, in comparison with isolated molecule fracture.

Consequently, any association must decrease chain tendency to degradation. However, it may be probable the existence of such intermediate particles at association, which possess lower height of the reaction barrier. In this case kinetic probabilities of the process performance increase. Sufficiently sharp increase of kinetic probabilities of the reaction must be observed in the case, if low molecular compound (oxygen, for example), participating in the reaction, is highly stressed. But it is necessary to remember, that even if kinetic probabilities of the process are increased, the reaction will proceed in the case of its thermodynamic benefit, also. As association depends on macromolecule concentration, it should be taken into account at the calculation of kinetic and thermodynamic parameters of the process, according to thermodynamics.

1.2. EXTREME SIZES OF CYCLIC COMPOSITIONS

Free energy of solution formation, containing cyclic large molecules, is expressed before degradation as follows:

$$G_{in}=G_x+G_{id}+G_{ex}+G_{defl}+G_s+\Delta G_s+G_{p-s}.$$

Free energy of solution formation, containing linear molecules, after degradation equals:

$$G_e=G_s-G_x+G_{id}+G'_{ex}+G'_{defl}+\Delta G'_s+G'_{p-s}+\Delta G_c.$$

The main symbols of both equations are similar to the ones which physico-chemical sense was explained in previous paragraph. ΔG_c - the change of free energy at chemical bond break in cycle and noncyclic bonds formation. G_{id} is equal for initial and end states.

It was supposed at consideration, that solvent molecules and macrochain molecules, including cyclic ones, are not associated with each other. It is supposed also that if solutions are diluted, we can neglect the contribution of the volume change into the change of process free energy. As the number of molecules changes at cycle break, the change of free energy of the process equals:

$$\Delta G = G_e - G_{in} = (G'_{ex} - G_{ex}) + (G'_{defl} - G_{defl}) + (\Delta G'_s - \Delta G_s) + (G'_{p-s} - G_{p-s}) + \Delta G_c \tag{15}$$

It is evident that all the abovementioned relates to the reaction of cycle opening type and ten formation of active centers (radicals) at molecule ends:

$$\rightarrow \bullet \sim\sim\sim\sim\sim\sim\sim\sim\sim\sim\sim\sim\sim \bullet$$

According to P. Flori approximation, G_{ex} and G'_{ex} equals respectively:

$$G_{ex} = -RT\left[\frac{N_1}{N_1+N_2}\cdot\ln\frac{N_1+N_2}{N_1+rN_2} + \frac{N_2}{N_1+N_2}\cdot\ln\frac{r(N_1+N_2)}{N_1+rN_2}\right]$$

$$G'_{ex} = -RT\left[\frac{N_1}{N_1+N_2}\cdot\ln\frac{N_1+N_2}{N_1+(r+t)N_2} + \frac{N_1}{N_1+N_2}\cdot\ln\frac{(r+t)(N_1+N_2)}{N_1+(t+r)N_2}\right]$$

Here t is the change of the number of contacts chain molecule-solvent at cycle break

$$\Delta G_{ex} = G_{ex}^{'} - G_{ex} = -RT\left[\ln\frac{N_1 + rN_2}{N_1 + (t+r)N_2} + \frac{N_2}{N_1 + N_2}\cdot\ln\frac{r+t}{r}\right] =$$

$$= -RT\left[\ln\left(1 - \frac{tN_2}{N_1 + (t+r)N_2}\right) + \frac{N_2}{N_1 + N_2}\ln\left(1 + \frac{t}{r}\right)\right].$$

(16)

In many practical cases t is sufficiently lower, than r, and the more so $\dfrac{tN_2}{N_1 + (t+r)N_2}$ is smaller than 1 in diluted solutions. So let expand ΔG_{ex} in series by x_2:

$$\Delta G_{ex} = -RT\left\{\left[\frac{tx_2}{x_1 + (t+r)x_2} + \frac{1}{2}\left(\frac{tx_2}{x_1 + (t+r)x_2}\right)^2 + \right.\right.$$

$$\left.\left. + \frac{1}{3}\left(\frac{tx_2}{x_1 + (t+r)x_2}\right)^3 + ...\right] + x_2\left[\frac{t}{2} - \frac{1}{2}\left(\frac{t}{2}\right)^2 + \frac{1}{3}\left(\frac{t}{2}\right)^3 + ...\right]\right\}$$

At very low concentrations we can neglect $tx_2/[1+(t+r-1)x_2]$ value, and the more so neglect all other terms of expansion $\ln\left(1 - \dfrac{tN_2}{N_1 + (t+r)N_2}\right)$.

In this case ΔG_{ex} decreases linearly with x_2 growth (increases by its absolute value). If $t<r$ (big cycles), one can restrict by the first term $ln(1+t/r)$, then $G_{ex}=-x_2(t/r)RT$. At the break of very big cycles at low concentrations (formation of infinitely long chains) the value of ΔG_{ex} approaches zero. It is more correct not to use the value of excessive free energy of mixing the molecules of different sizes, calculated from I. Progojin's model [19], for thermodynamic estimation of the probability of relatively small size cycle break.

If cycles being broken are relatively large, $(G_{p-s}^{'}-G_{p-s})$ and $(\Delta G_s^{'}-\Delta G_s)$ will be close to analog values at the break of low molecular compounds. As the change of polymer-solvent contact number will depend weakly on molecular weight of the cycle, then both $(G_{p-s}^{'}-G_{p-s})$ and $(\Delta G_s^{'}-\Delta G_s)$ will be constant. At relatively small sizes of the cycle the number of macrochain molecule-solvent contacts can be connected nonlinearly with the chain length. Then, as in previous paragraph,

$$\Delta G = C_{ch}^{+}(G_{defl}^{'} - G_{defl}) + (G_{ex}^{'} - G_{ex}),$$

where $C_{ch}^{+} = \Delta G_c + (G_{p-s}^{'} - G_{p-s}) - (\Delta G_s^{'} - \Delta G_s).$

As ΔG_{ex} value in diluted solution is small, in a number of practical cases

$$\Delta G = C_{ch}^+ + (G'_{defl} - G_{defl}).$$ (17)

The term (G'$_{defl}$-G$_{defl}$) characterize labor, spent for the displacement of linear chain ends to the distance equal to the length of forming bond. Mean-square of the distance between linear chain ends increases with molecular weight growth. It is clear, that average distance between chain ends will increase also. Evidently, labor required for chain ends transferring will increase with molecular weight growth, and a moment will occur, at which this labor gain will exceed free energy of chemical bond formation. Thermodynamic opportunity of cycle break appears at $G'_{defl} - G_{defl} = C_{ch}^+$. Term (G'$_{defl}$-G$_{defl}$) possesses entropic character, mainly. Unlike chain degradation reaction critical length of thermodynamically stable cycle will be observed for isolated chains, as well as for the chain in solution, the better solvent is from the point of view of chain hardening, at lower molecular weights thermodynamic opportunity of cycle break will be displayed. One can mind, that the chain length, at which cyclization reaction is still thermodynamically probable, is at any rate not greater than the one, at which linear chain becomes thermodynamically instable, i.e. it can be treated by degradation. Apparently, if one of chemical bonds of the chain is weaker, it cannot be excluded the possibility of its stretching with cycle size growth, and cycle break will occur at this bond. If all chemical bonds in the cycle are uniform, then the stretch of each bond will be lower, than if similar bond would be the weakest in the cycle.

As the change of free energy is uniform for the bond break in infinitely long chains and low molecular compounds, its value can be determined not only by the break of the largest cycle, then. As this value is constant in the first approximation, then the variation of free energy change of the recyclization process will define the change of free energy by the account of system flexibility change, i.e. it will define the increase of labor, required for chain ends transfer from particular distance to the one, equal to chemical bond length (equation 17). The distribution by bond lengths may not be taken into account at the analysis, because fluctuation of bond lengths is low, comparing with that of the distances between chain ends. If G$_{defl}$ value can be estimated according to experimental data or on the basis of corresponding model, then determining C$_{ch}$ and molecular weight, at which the cycle becomes instable thermodynamically, one can calculate G'$_{defl}$, i.e. free energy of cyclic compound deflection. For the chain, consisting with one conformer in the solvent, it is possible to estimate G$_{defl}$ according to [12], and for isolated chain - according to [14]. Maximum value of deflection free energy of isolated chain equals nRT by its absolute value, and minimum value of free energy of cycle deflection equals zero (the analog of "rigid bar"). That is why $C_{ch} \geq nRT$. Typical example of bond weakening in the cycle (precisely the increase of its rigidity or flexibility) is complex formation [21]. In presence of complex formation at the increase of cycle rigidity its break can be observed (in absence of complex formation the cycle can be thermodynamically stable). We have not considered the question about the influence of rotary isomery on the process of cycle degradation in the present case. At temperature growth G'_{defl} increases, approaching the limit, defined by nRT expression (the chain approaches "free rotation of bonds"). As the length of breaking bond changes insufficiently, then $(G'_{defl}-G_{defl})$ approaches nRT according to temperature growth. In general case the value of C_{ch} decreases with the temperature according to thermodynamics. The deviation of the observed dependences according to the temperature from similar ones of the processes, in which only low molecular compounds participate, can be stipulated by the change of chain-solvent contacts according to temperature.

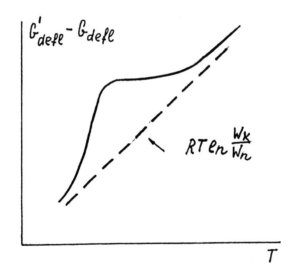

Figure 15. The dependence of deflection free energy changing during chain scission $(G'_{defl}-G_{defl})$ on temperature.

If C_{ch} decreases, and $(G'_{defl}-G_{defl})$ increases with the temperature, minimum size of thermodynamically stable cycle decreases then. If C_{ch} increases, that may be observed very rare, then maximum size of thermodynamically stable cycle may depend on the extreme temperature. More complicated temperature dependences of maximum size of thermodynamically stable cycle are probable, also. Extreme size of thermodynamically stable cycle is connected linearly with the temperature, at which $(G'_{defl}-G_{defl})/nRT$ changes weakly (very low or very high temperatures), and C_{ch} changes in linear order (according to approximations of ΔH and ΔS constancy in narrow temperature ranges, accepted in chemical thermodynamics). Let admit, that all chemical bonds are uniform before break, or one of them differs from the rest (Naturally, the formation of more complicated chemical composition of cycles is possible too). Chemical bonds in chains will differ from the end groups after cycle break. That is why G'_{defl} of linear chains possesses uniform temperature dependences in both cases. For initial cycles, containing two types of bonds, it should be taken into account two values of local rigidities and parameters of chain-solvent interactions. Temperature dependences of G'_{defl} and G_{defl} for constant n value, i.e. in absence of chemical processes in cycles and linear chains, are shown on Figure 15. Consequently, in this case $(S'_{defl}-S_{defl})$ will possess extremal character (extremum is weakly expressed). The probability of the formation of cycles with small amount of bonds is stipulated by free energy gain at bond formation as well as by its loose in consequence of valent angles deformation (of bond lengths rarely). It is described in detail in many manuals on theoretical organic chemistry (see for example [22]). Thus there may be observed two borders of thermodynamically stable cycle existence in dependence on the chain length.

As the size of thermodynamically stable cycle depends preferably on flexibility change during the process, it will depend on its disposition in chain molecule, too. To simplify the consideration let us take a new cycle, contained in chain molecule (Figure 16). The formation of cycle in the chain makes the latter more rigid usually (local rigidity of cycle disposition place in the chain is higher, than that of the rest of the chain). Estimating G'_{defl}, one can consider a chain, containing three parts (two parts, if cycle is disposed at the end of the chain), two of which (end ones) possess uniform local rigidity. The dependence of $\overline{R}^2$ for such chains in solvent and in isolated state is shown on Figure 17. According to Figure 17, chains possess higher formation free energy, when the cycle is disposed in the middle of the chain or in the chain points, at which $\overline{R}^2$ is maximum (the smaller $\overline{R}^2$ is, the higher chain entropy is). After completing the reaction local rigidity of the chain part, where the break has occurred, can be even lower than that of the

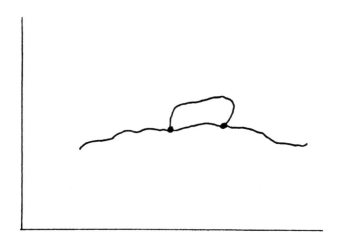

Figure 16. Scheme of cycle disposition in the chain.

rest of the chain parts. So maximum value of $(G'_{defl}-G_{defl})$ will depend on the place of initial cycle disposition. At close influence of solvent on local rigidities of various parts $(G'_{defl}-G_{defl})$ will possess extremal character in dependence on the reaction place (Figure 18): maximum value will be observed, when the initial cycle is disposed in the middle of the chain, and minimum one - at the end. As the change of energy at cycle break and new bond formation depends weakly on the chain length, then extreme size of thermodynamically stable cycles will depend extremely on the place of disposition in the chain (maximum size of thermodynamically stable cycle will be observed at chain ends). It is clear that the solvent may change the present dependence (Figures 12-14).

At the increase of the number of cycles in the chain free energy of deflection in initial and end states will increase; however $(G'_{defl}-G_{defl})$ will change extremely, i.e. the cycles will be more stable, disposing at the end of the chain. Cycle break is supposed in the present example. Maximum of $(G'_{defl}-G_{defl})$ absolute value will increase with local rigidity growth of the main chain as well as of the cycle. Consequently, the break of the cycle disposed in the middle of the

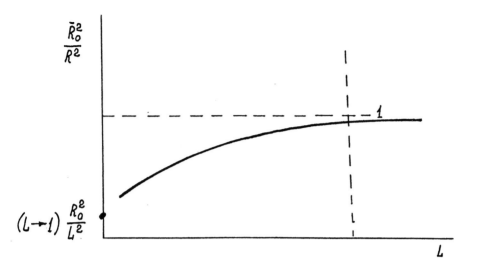

Figure 17. The dependence of $\overline{R}_0^2 / \overline{R}^2$ on length chain L.

chain, will be the most profitable, the more rigid the chain is, and, consequently, the more cycles dispose in it. Minimum $(G'_{\text{defl}}\text{-}G_{\text{defl}})$ absolute value will depend weakly on chain rigidity, including the number of cycles in it, because $\overline{R}^2$ change at the change of local rigidity of the chain ends varies weakly. On the whole, the dependence of thermodynamically stable cycle size on conditions (temperature, solvent, etc.) and on chain structure (local rigidities, length) is defined by "chain effect" [23], i.e. by entropy change at variation of the chain deflection in all cases. Thus, the solvent, interacting more strong (stretching) with the chain, can change sufficiently the place of more stable cycle disposition in the chain (Figures 12-14). As with the chain length growth there increases the rigidity of chains, containing cycles, as well as the chains after their break, then $(G'_{\text{defl}}\text{-}G_{\text{defl}})$ value decreases. For infinitely long chains $(G'_{\text{defl}}\text{-}G_{\text{defl}})$ change becomes minimal, and the sizes of thermodynamically stable cycles are determined in analog with the case, in which the cycle is formed at chain ends. It is clear, that the size of thermodynamically stable cycle is leveled with the chain growth. Flexibility of initial and end chain increases according to temperature growth, that leads to leveling of the extreme size of

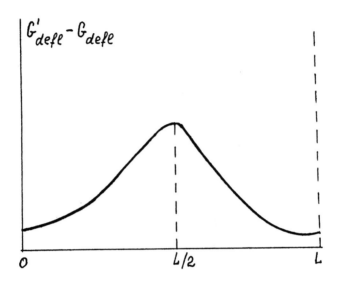

Figure 18. The dependence of deflection free energy changing on of reacted place disposition in the chain.

thermodynamically stable cycle. However, in many cases this size decreases in consequence of parameter C_{ch} change, which value, as mentioned above, does not depend on cycle disposition in the chain, and depends on process conditions, only. If chemical bonds in the cycle are uniform, then cycle break proceeds most probably by the place, where both parts of side chain, formed at the break, possess maximum size (Figure 19). If one or several chemical bonds are disposed in the main chain, the most probable place of the break in the chain will be defined by the change of distances between ends and, probably, will depend strictly on reaction conditions and chain structure. If the break occurred by the bond, disposed in the chain, then there occurs the change not only of local rigidity, but the increase of its contour length, too, and consequently, of observable flexibility. The break, occurred by bonds, disposed in side chain, increases local rigidity of the reaction place without changing contour length of the main chain. Correlation of these contributions into the change of deflection free energy defines the most probable place of the reaction and, consequently, the place, at which the reaction proceeding becomes thermodynamically probable, reaching corresponding conditions. At temperature increase as well as at the increase of the main chain length, the break in various places of cycles approaches equally probable one. It should be mentioned that different stability of various cycle sizes may lead to various cycle compounds (cycles), formed, for example, by complex formation, able to participate in other processes further on, for example, in degradation and catalysis [20]. The influence of solvent on the most probable place of the reaction proceeding is defined by its interaction with local places of the chain and is corresponded strictly to "chain effect," shown in [13].

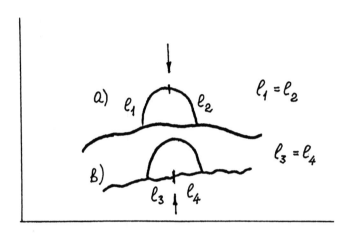

Figure 19. Scheme of most probable place of cycle break in the chain. a - the break occurs by the bond of cycle. b - the break occurs by the bond of chain itself.

1.3. EXTREME SIZES OF CHAIN PARTS BETWEEN NETWORK POINTS

Free energy of network formation before the first break is:

$$G_{in} = G_x + \sum_i G_{defl,i} + G_{im} + A + G_{im} + G_{p-s} + \Delta G_{s-s} + G_s$$

Here G_{im} - free energy of intermolecular bond formation; $\sum_i G_{defl,i}$ - total free energy of nonassociated chain parts deflection (it is supposed that free energy of associated parts deflection is close to zero); A - labor, performed over the system or by it; G_s - free energy of solvent formation, disposing in the network or contacting with it; G_{com} - free energy, stipulated by combinatorial entropy of solvent molecule disposition in the network.

After the break of a single bond in the system free energy equals

$$G_e = G_x + \sum_j G_{defl,j} + G'_{im} + A' + G_s + \Delta G'_{s-s} + G'_{p-s} + G'_{com} + \Delta G_x$$

where (') index marks the values, related to the end state; ΔG_x - the change of free energy at network break and formation of new bonds. As in the above described cases, we neglect the change of system volume and conformation change

$$\Delta G = G_e - G_{in} = (\sum_j G_{defl,j} - \sum_i G_{defl,i}) + (G'_{im} - G_{im}) + (A' - A) +$$

$$+ (\Delta G'_{s-s} - \Delta G_{s-s}) + (G'_{p-s} - G_{p-s}) + (G'_{com} - G_{com}) + \Delta G_x \qquad (18)$$

In fact, the following model was considered at writing thermodynamic equation [24]. Thermodynamic system contains 1 mole of elements of uniform structure, each of them representing a part of chain between chemical points. Each part contains n_x bonds of uniform chemical structure. We neglect differences in chemical structure of the point. Stress, which is induced by internet points is calculated in $\sum_j G_{defl,j}$ and $\sum_i G_{defl,i}$, and in A' and A also. Chain parts between cross-linkages may form intermolecular (physical) bonds, the amount of which is n_{ph} per element, and free energy of formation is G_{im}. If it is supposed that single chain of a part can form one intermolecular bond with a chain of opposite part, then $n_x \geq n_{ph} \geq 0$. Labor is attached to network elements. This labor can be performed by means of external influence, as well as by means of "internal" forces, occurred in the sample during production, storage and exploitation of the article. Moreover, there exists the solvent in the network in consequence of swelling. It stretches chain parts. If the network is swelling in a good solvent, n_{ph} approaches zero in the limit. Consequently, limitly swelled polymer is considered here. Its swelling scantiness is stipulated by both chemical and physical points.

As in previous cases, the state is accepted as standard one, in which elements are stretched completely ("rigid here"). Solvent amount in the system is constant. As we consider just the initial break, the number of molecules does not change practically then, and $(G'_{com} - G_{com})$ is close to zero, i.e. the role of combinatorial term in swelled network, which may be dominant in the reactions in solutions, is insufficient. The term $\Delta G_{com} + (G'_{p-s} - G_{p-s}) + (\Delta G'_s - \Delta G_s)$ may be accepted constant then - C_c. Under more accurate consideration the number of solvent molecules will be defined by sizes of chains and solvent molecules, as well as by the amount and distribution of intermolecular bonds in the parts. The present supposition is correct for the case, if n_x is high enough, and $(n'_{ph} - n_{ph})$ number is changed by constant value. Then

$$\Delta G = C_c + (\sum_j G_{defl,j} - \sum_i G_{defl,i}) + (A' - A) + (G'_{im} - G_{im}). \qquad (19)$$

Consequently, ΔG is defined by C_c coefficient as well as by the change of element deflection, labor over the system and the number of intermolecular bonds. The value of C_c approaches ΔG value, observed in similar reactions with the participation of low molecular compounds, only. As intermolecular bonds are distributed in elements according to Gibbs distribution, then chain parts between molecular bonds and branching points possess different lengths, in which the lengths of nonassociated parts are different, also. Gibbs distribution is performed in polymer equilibrium, only, which usually exist in so-called stationary states.
Then

$$\sum_i G_{defl,i} = (n_x - n_{ph})RT\alpha$$

where α - coefficient, characterizing a part of free energy of deflection, rested from free energy at free rotation of bonds, fitted single bond. It is clear, that in general case α is the averaged value, taking into account the sizes of nonassociated parts and their disposition in the element. $\sum_i G_{defl,i}$ characterizes averaging deflection free energy by sizes of nonassociated parts. In analog

$$\sum_j G_{defl,j} = \alpha'(n_x - n_{ph})RT.$$

Then we obtain

$$\sum_j G_{defl,j} - \sum_i G_{defl,i} = RT[\alpha'(n_x\ell - n_{ph}) - \alpha(n_x - n_{ph}). \qquad (20)$$

As free energy of deflection decreases total free energy, then it must enter the expression with negative sign. According to accepted approximation G_{im} is defined by multiplication of free energy and the number of bonds, because there is used the principle, similar to P. Flori one for chemical bonds. The energy of intermolecular interaction $E_{ph}=\beta RT_b$ [25], where T_b - boundary temperature, over which we can neglect intermolecular interaction (entropy losses are calculated by free energy of deflection); β - coefficient, characterizing the part of heat energy, accepted by the bond, $1>\beta>0$. Then

$$G_{im} = n_{ph}\beta RT_b; \; G'_{im} = \beta n_{ph}RT_b; \; G'_{im} - G_{im} = \beta RT_b(n'_{ph} - n_p). \quad (21)$$

Free energy of intermolecular interaction decreases also general free energy of system formation. Last component of free energy change in the process is labor change. Labor can be positive and negative, according to the type and value of the stress. At swelling there occurs the stretch of nonassociated parts of the chain, leading to the decrease of their flexibility. That is why swelling is equivalent to stretching load.

Equation (19) may be rewritten as follows:

$$\Delta G = C_c - RT[\alpha'(n_x - l - n'_{ph}) - \alpha(n_x - n_{ph})$$
$$-\beta RT_b(n'_{ph} - n_{ph}) + (A' - A) \quad (22)$$

As it is seen from the equation (22), thermodynamic opportunity of the reaction initiation ($\Delta G=0$) is defined by network properties (C_c, n_x, T_b), as well as by conditions of production, storage and exploitation (n_{ph}, α), and by external influence (T, A), also. As it was mentioned above, C_c value is the function of chemical structure of the network and the solvent (in a number of cases solvent amount, disposed in the network, may depend on n_{ph} and intermolecular bonds distribution).

If the system is not swelled ($[\Delta G'_{s-s}-\Delta G_{s-s}]=0$; $[G'_{p-s}-G_{p-s}]=0$) and there is no labor, influencing it, then $C_c=\Delta G_{com}$, and the temperature, at which thermodynamic opportunity of reaction proceeding occurs, is determined from correlation of n_{ph} and α:

$$T = \frac{\Delta H_x + \beta RT_b(n_{ph} - n'_{ph})}{\Delta S_x + [\alpha'(n_x - l - n'_{ph}) - \alpha(n_x - n_{ph})]}. \quad (23)$$

The present conditions characterize material storing in the initial approximation. It is natural, that network structure must be stated.

If polymer system would be able to exist in equilibrium state only, then strictly defined correlation between (α, n_{ph}) and (α', n'_{ph}) would exist in particular conditions, according to minimum of free energy of system formation. Consequently, there would occur one temperature only, at which process initiation is thermodynamically probable. In very rare cases there may occur different correlations between (n_{ph}, α) and (n'_{ph}, α'), which display one and the same value of free energy minimum of system formation.

It is known that polymers may exist in various stationary states, which are defined by the amount and distribution of intermolecular bonds in the sample at definite network structure. The latters are defined by the conditions of storage, exploitation and production of the network. That is why T values may be different. The highest value is observed in the case of equilibrium state of the system. In this case it is necessary to point out, that n'_{ph} value becomes close to n_{ph} one at n_x, great enough.

There are many proofs in literature of chemical reaction initiation in the network, starting from definite temperature. Thus, it was observed at the investigation of fracturing cross-linked butadiene copolymers with acrylonitrile, that below some temperature observed constant rate of one of degradation reactions was several decimal degrees lower, than that estimated according to Arrhenius equation [26], although linear dependence of Arrhenius equation anamorphase was observed till definite temperature (rate constant approached zero abruptly with the temperature decrease). If polymer existed in highly elastic state, then it is necessary to take into account both intermolecular bonds and flexibility of chain parts. If polymers exist in vitrious state (plastics), the contribution of deflection entropy is close to zero, and system stability increases according to the equation (23), i.e. it increases with the growth of the number of intermolecular bonds. Estimating the influence of n_{ph} and α on system stability, one should take into account, that $0<\alpha<1$, and n_{ph} may change in very wide ranges ($0 \leq n_{ph} \leq n_x$). If there are no intermolecular bonds in the system, we can determine temperature value, at which process proceeding becomes probable, as follows:

$$T = \frac{\Delta H_x}{\Delta S_x + (\alpha' - \alpha)n_x - \alpha' l}.$$

We can neglect values of $(\alpha'-\alpha)$ and $\alpha' l$ for sufficiently long chain parts between branching points. In this case E value should not differ from the analog one for the process with the participation of low molecular compounds, only. In the rest of cases T value is lower, i.e. T dependence on n_{ph} possesses extremum character (the minimum exists). These circumstances, namely, define sufficiently high chemical stability of low molecular compounds, comparing with polymers, possessing similar chemical structure. The load, attached to the network, can stretch non associated parts of chains between branching points or intermolecular bonds, so it promotes the increase of the number of intermolecular bonds (polymer network transites to vitrious state from highly elastic one) It also can create the stress on intermolecular bonds, decreasing T_b (stretching them), change valent angles and lengths of chemical bonds. As the labor, necessary for these actions, possesses different values, one of the effects can be predominant at various stress values. As in general case under deformation of valent angles and bond lengths stress value is more sufficient, than at the change of flexibility and molecular bonds deformation, the influence of it on ΔH_x and ΔS_e is displayed at relatively large values, only. As stress contribution into deformation of intermolecular bonds is usually higher, than into flexibility change, then numerator increases and denominator decreases according to stress attachment. Consequently, at low stresses according to the increase of intermolecular bond number, there increases the temperature, at which the process initiation becomes probable. Further on this temperature begins to decrease. The most abrupt decrease of it will occur at valent angle deformation and chemical bond lengths, then. It is necessary to point out, that in present consideration it is supposed, that the load influences sufficiently higher the initial state, instead of the end one. The present supposition is stipulated by the circumstances, that in the initial state the ends of chain parts are influenced by stretching forces, and in the end state this force decreases at least (the stress transites to neighbor parts)   .

Stress occurring during production, storage and exploitation of the sample is explained by this very fact, that one and the same temperature of probable reaction may be obtained at corresponding stationary state as well as at stress attachment to equilibrium state. That is why at low stresses there may proceed the transition from one state to another, accompanied by the increase of system entropy, and then there will be observed its decrease. Energy change according to the change of intermolecular bond number at system entropy increase under the stress influence is compensated by flexibility increase. The value of free energy change, depending on the decrease of intermolecular bonds amount, becomes higher, than that, connected with flexibility decrease, after reaching definite stationary state, only. Then the number of intermolecular bonds starts increasing.

All abovementioned relates to stretching stress. However, we can consider similarly the other ones, including more complicated stresses.

The solvent, in which polymer network swelled, is able to change the number of intermolecular bonds, to decrease observable flexibility of chain parts between points and to stretch the system, i.e. to perform labor over it. Its influence on C_c parameter, according to accepted approximations, is similar to the influence on the process, in which low molecular compounds participate, only. Some deviation can be observed in consequence of solvent molecules supressing in the network. One can mind, that the change of observable local rigidity of the chain in consequence of network swelling is similar by its first approximation to its change for analog chains in solutions. In some cases the number of solvent molecules, disposed in the network at its swelling, will depend on both chemical structure of the chain and the solvent and the number and distribution of intermolecular bonds in the sample, in consequence of observable local rigidity dependence on the length of nonassociated parts. For example, if it is necessary to break single intermolecular bond, disposed between associated and nonassociated parts, the change of C_c parameter will be uniform in all cases, and the gain in deflection free energy will be different. Consequently maximum amount of the solvent in the network will depend on both chemical structure of the compositions and on sample prehistory. If solvent amount is lower, than maximally possible one (incomplete swelling), it will be disposed in places, where it is possible not to change intermolecular interaction and where flexibility of chain parts will change minimally (at chain stretching) and maximally in rare cases of chain supressing to the value, at which observable local rigidity is close to free rotation. Consequently, the solvent will be distributed in the sample nonequally probable. First of all, it will be distributed between nonassociated parts of chains of optimum size (in analog to interphase plastification). Then it will dispose in the zones, in which free energy of deflection will grow, according to the increase (in equilibrium state). At the end, if it is possible, it will penetrate into other zones, breaking more and more bonds.

At the attachment of the stress, lower and it is required for deformation of valent angles and chemical bond lengths, in general case maximal amount of the solvent in the network should decrease. It is evident, that if the number of intermolecular bonds and their distribution decreases, temperature should decrease (if C_c=const), and at their break temperature will decrease more. In present consideration it was accepted also, that the influence of the solvent on the initial state is higher, than on the end one (in analog to stress influence). The increase of particular temperature can occur as a consequence of C_c decrease and at transition from one stationary state to another.

1.4. EXTREME SIZES OF CHAINS ON THE SURFACE

Free energy of the formation of the disposed on the surface is:

$$G_{in} = G_x + G_{id} + G_c + \sum_i G_{defl,i} + G_m + G_n + \Delta G_{mc}$$

Free energy of chain formation on the surface after degradation is:

$$G_e = G_x + G_{id}' + G_c' + \sum_j G_{defl,j} + G_m' + G_n + \Delta G_x + G_{mc}'$$

Here ΔG_x - free energy of chain break and formation of new bonds; G_m - free energy of chain-surface bond formation; G_c - excessive combinatorial free energy, stipulated by different disposition of chain molecules on the surface; ΔG_{mc} - combinatorial free energy, stipulated by different disposition of intermolecular chain-surface bonds on chain molecule. The rest of G terms possess the above-mentioned physical sense. Index (') relates to the end state of the system.

$$\Delta G = G_e - G_{in} = (G_{id}' - G_{id}) + (G_c' - G_c) + (\sum_j G_{defl,j} - \sum_i G_{defl,i}) + \tag{24}$$

$$+(G_m' - G_m) + (G_{mc}' - G_{mc}) + \Delta G_x.$$

We neglect the contribution into free energy of the initial and the end state, depending on the volume change. It is supposed, that the degree of surface filling is low (in analog to diluted solution). As at the consideration in the previous paragraph we will accept, that local rigidity of parts with intermolecular bond is nearly infinite. In consequence of the distribution of intermolecular bonds along the chain and between them we use ΣG_{defl} value. In this case

$$\sum_i G_{defl,i} = \alpha(n_x - n_{ph})RT; \quad \sum_j G_{defl,j} = \alpha'(n_x - \ell - n_{ph}')RT$$

As in previous case, general number of adsorbed initial molecules is 1 mole. If intermolecular bonds are great enough, molecules in the volume cannot transite, and there is no motion at the surface. In this case we can neglect terms, connected with combinatorial entropy of molecules disposition in the volume, and the more so their difference:

$$\Delta G = \Delta G_x + (\sum_j G_{defl,j} - \sum_i G_{defl,i}) + (G_{im}' - G_{im}) + (G_{mc}' - G_{mc})$$

In analog to the previous paragraph:

$$G_{im}' = \beta n_{ph}' RT_b; \quad G_{im} = \beta n_{ph} RT_b.$$

As it is usual, n_{ph} represents mean-arithmetical number of intermolecular bonds, n'_{ph} being total amount of bonds in both parts of fractured initial molecules:

$$G'_{im} - G_{im} = \beta R T_b (n'_{ph} - n_{ph}).$$

Terms G'_{mc} and G_{mc} are defined by combinatorial entropy, depending on transposition of intermolecular bonds at the molecule, taking into account geometry of the surface and the chain (some intermolecular bonds, for example, the ones at neighbor chain atoms cannot be performed in consequence of the structure of the chain itself or because of group disposing, which form these bonds on the surface). So thermodynamic opportunity of the initiation of the degradation reaction, as well as any other process will be defined by properties of the surface and chain features. In the simplest case intermolecular bonds can be formed at any place of the volume - combinatorial term is defined by the totality of transposition number of a definite bond amount from 0 to the formation of intermolecular bonds on all active centers. This totality equals Newton binomial. According to the definition, the formula can be obtained in the case of equal probability of the formation of intermolecular bonds. In real case the number of states is stipulated by chemical and geometrical composition of the chain and the surface and can be sufficiently lower. Then

$$G'_{mc} = n' R T_{ex} + n'' R T_{ex}; G_{mc} = n R T_{ex}$$

Here n' and n'' - the amount of bonds, formed in molecules.

$$G'_{mc} - G_{mc} = l R T_b.$$

If l is small, comparing with n, then we can neglect this term. In this case:

$$\Delta G = C - \beta R T_b (n'_{ph} - n_{ph}) - [\alpha'(n_x - n'_{ph} - l) - \alpha(n_x - n_{ph})] R T. \quad (25)$$

Concluding from the last expression, temperature, at which reaction initiation is thermodynamically probable, is determined from the equation:

$$T = \frac{\Delta H_x + \beta R T_b (n'_{ph} - n_{ph})}{\Delta S_x + R[\alpha'(n_x - n_{ph} - l) - \alpha(n_x - n_{ph})]} \quad (26)$$

Consequently, at definite number and distribution of intermolecular bonds degradation process may begin at definite temperature only, if there are kinetic probabilities for the reaction proceeding at the present temperature. At lower temperature n_{ph} increases and α decreases, i.e. numerator does not change at least (grows more frequently), and denominator decreases, approaches ΔS_c. Moreover it is necessary to point out, that at low temperature n'_{ph} is also high. That is why it is hard to remove formed chain fragments, that specially promotes their combination at radical break mechanism. If molecule break occurs at the end of the chain, then formed molecule can be removed into space. This promotes the break at relatively low temperatures. In the limit $n_{ph}-n'_{ph}=1$. At higher temperatures $(n_{ph}-n'_{ph})>1$, so the break in the chain middle becomes probable (denominator increases). The case promotes the reaction proceeding at chain ends at low temperatures, that at uniform length of nonassociated part it is more profitable thermodynamically for it to dispose at the end (number of states and, consequently, entropy will be higher for uniform length of nonassociated parts, if the part is fixed by one end, and not by both). We will consider this question below more in detail.

Considering the influence of various parameters on T, it is necessary to take into account, that α, n_{ph}, α', n'_{ph} depend on n_x and on each other. Parameters α and α' increase with the chain length at constant n_{ph} and n'_{ph} and decrease with n_{ph} and n'_{ph} growth at constant nx value. As the change of α lies in relatively narrow ranges comparing with n_{ph}, in the most number of cases predominant influence on the degradation process is performed by the amount and distribution of intermolecular bonds, necessary for the chain retaining on the surface, which is usually lower, than maximally possible one, and the difference between them increases with the chain length growth, then the association of molecules on the definite surface with different amount and distribution of these bonds is principally possible. It follows from this fact, that the probability of the equilibrium formation, as well as of stationary states at adsorption, occurs in dependence on storage conditions. Apparently, different amount and distribution of intermolecular bonds is one of the main causes of stationary state formation.

If $n_{ph}-n'_{ph}$ is (for example, molecule was fixed by both ends before the process, and then each fragment is fixed by one end), then α' is greater than α (flexibility of the chain, fixed by both ends, is usually lower, than that of chains, fixed by one end). In this case denominator increases and T decreases with n_x growth. Though it is clear, that α grows with n_x, also, molecule can be instable at any definite temperature at definite length. If the chain is stretched sufficiently by the surface (intermolecular bonds are removed from each other sufficiently far), then we can accept for the first approximation $\alpha \to 0$ and $\alpha' \to 1$. Then

$$T = \frac{\Delta H_x}{\Delta S_x + R(n_x - 3)}$$

At definite values of ΔH_e and ΔS_e from the equation the length of nonassociated parts is determined, at which degradation reaction is thermodynamically probable, T being lower than the temperature, at which the break of chemical bonds in low molecular compound proceeds. The value of α at definite n_x will be defined by the distance between engagement points (there exists optimal value of this distance, at which α possesses maximum value). The value of T will depend on engagement places, and apparently, on disposition and distribution of surface active centers. In the limit $\Delta H_x - T\Delta S_x = RT_{nx}$, where n_x for the first approximation equals the length of nonassociated parts of the chain. In this case it is necessary to take into account, that if $T > T_b$ (T_b-boiling temperature), molecules will evaporate, and if $T < T_b$, molecules will be decomposed until there form the chain fragments, being able to evaporate. Analyzing degradation it is necessary to take into account, that at temperature increase the value of free energy change of bond break decreases, and the difference of $\sum_j G_{defl,j}$ and $\sum_i G_{defl,i}$ increases.

The first term may increase in consequence of the growth of α' and T, and the second one - in consequence of temperature growth, only.

Considering the process proceeding on the surface, it should be taken into account the process of chemical degradation itself as well as the break of intermolecular bonds. However, at break of all intermolecular bonds there exists extreme length of thermodynamically stable chain on the surface, that is stipulated by impossibility of large molecule evaporation. Thermodynamic instability is defined by the fact, that if molecule occurs on the surface, there cannot be realized all the states in it, which would occur in gas. As the transition of molecules into gas phase depends on the attraction force, extreme size of thermodynamically stable chain will change in dependence on this attraction. In absence of the attraction molecules are able to transite into the gas, and their stability will be the same even for infinitely large molecules as for low molecular compounds. This can be easily shown by modelling isolated molecules. If molecule is fixed by two points on the surface, it represents something like cyclic compositions. In this case system stability depends on fixing places of the chain itself on the surface (Figure 20). The disposition will be mostly stable, when engagement places are situated nearby and closer to the chain end. Then the system may be considered as fixed by single end. Free energy of formulation will decrease according to translocation of fixation places, because longer ends possess high α value (for infinitely long chains α does not practically depend on their length). At the decrease of chain ends, i.e. broadening engagement places, free energy of the initial system formation will decrease and will become minimal, when both ends are fixed, disposed from each other at the distance, equal to contour chain length. In this state α will be close to zero, and α' will be sufficiently higher (in the limit α'-1). As two ends are formed at the chain break, and their free energy of deflection depends on length, then the break will occur at the middle of "cycle" according to the highest probability (Figure 21). If changes of free energy of chain-surface bond formation are lower, than that of chemical bond formation, then with regard to higher free energy of larger parts, fixed by one end, chain surface bonds will break with higher probability. This phenomenon must be observed at the association with the formation of physical bonds.

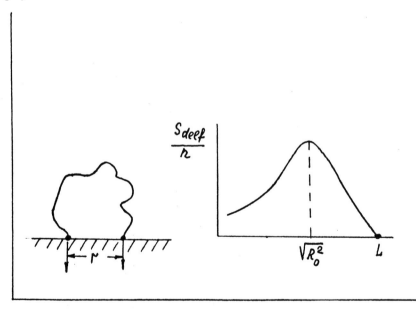

Figure 20. The dependence of deflection entropy related to monomeric unit (bond) S_{defl}/n on the distance of fixed points. The molecule is fixed by two points on the surface.

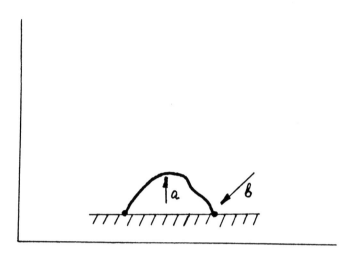

Figure 21. Scheme of probable break of molecule. Molecule is fixed by two points on the surface. Arrows show the break direction a - in the middle of cycle. b - on chain-surface bond.

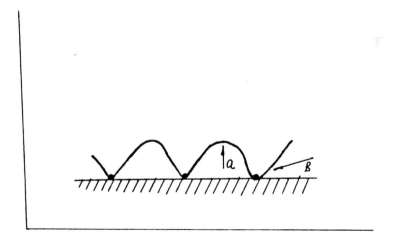

Figure 22. Scheme of adsorbed molecule break. Arrows show direction of break: a - break of chain; b - detaching of end of chain-surface bond.

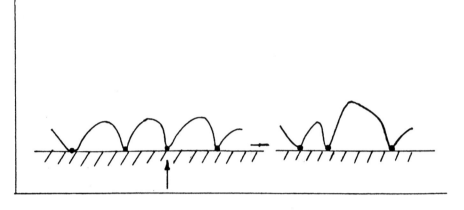

Figure 23. Scheme of detaching of chain-surface bond with increasing of length between two fixed points.

If molecule is fixed on the surface by three and more active centers, and if the energy of chain-surface bond is higher, than bond energy in macromolecule (chemical chain-surface bond), degradation may be considered for different cycles separately. In this case the break will by profitable thermodynamically with higher probability, occurring at the middle of the cycle. If the bond energy in the chain is higher, than that of chain surface bond, then reaction initiation is more probable, proceeding by chain-surface bond. In this case reactions, forming longer nonassociated ends, are preferable, i.e. the break of the "end" chain-surface bonds is more profitable thermodynamically (Figure 22). However, in the case of probable formation of large cycle (Figure 23) the break of chain-surface bond will be profitable thermodynamically for "internal" bonds, too, in consequence of high benefit of free energy by means of the increase of nonassociated part length in the cycle, comparing with the increase of the end part length. For example, if the length of the end part increases by one unit, and large cycle is formed at the break of middle cycle, the reaction of cycle formation is thermodynamically preferable. The place of reaction proceeding depends on entropic components of free energy, mainly. Both reaction places are equally probable, if free energy change is the same for equal number of breaks of variously disposed chain-surface bonds. In this case equal probability of the break may be observed at the reactions of cycle break and at the growth of the chain end, and at the reactions of cycle size increase. If lengths increase equally (for example, such case may be displayed at equal probability of bond formation at any place of the chain), the increase of nonassociated end length is preferable always. That is why chain ends are less associated, than its middle at adsorption, macromolecule association and other similar processes. Calculation of combinatorial free energy change should be performed in presence of relatively weak (physical, preferably) chain-surface bonds. The influence of this component change on the process is similar to that of excessive combinatorial entropy of different size molecules in solution (see paragraph 1). However, absolute value of excessive combinatorial entropy on the surface is usually lower, than for the solution, because the number of various dispositions of molecules in the solution is calculated for three dimensions, and on the surface - for two. Solvent influence on the change of process free energy is displayed by practically all the terms, except $(\Delta G'_{mc} - \Delta G_{mc})$. Thus, estimating $(G'_c - G_c)$, it is necessary to take into account combinatorial analysis, too, in consequence of the surrounding of nonassociated parts of chain molecules and contacting with nonassociated molecules. As it was mentioned above, the solvent makes the chain more rigid in the most number of cases. That is why in many cases $(\sum_j G_{defl,j} - \sum_i G_{defl,i})$ value can decrease.

Moreover, the solvent changes local free energy of chain-surface bond formation, because at its break there may occur the gain of free energy of surface formation. In the presence of solvent in the first approximation it may be taken, that all local places of the surface, able to form intermolecular bonds, are occupied by these bonds or by solvent molecules. That is why $(G'_{id} - G_{id})$ will change weakly in comparison with their value in solvent absence, if it is considered the disposition of molecules on the surface, only. It will depend also on system volume, as well as on the surface size, if general combinatorial analysis of the system is calculated with regard to solvent molecules, disposed on the surface. however in the last case it is necessary to take into account transfer of initial and degradated polymer molecules and solvent ones to the volume (into the solution). Moreover, it is necessary to introduce components of free energy change, connected with the change of local interaction of the molecules, surrounding the chain, with each other, and local interaction of associated and nonassociated parts of chains with the solvent. The influence of these components is similar to their influence on molecule degradation in the solution. Only $(G'_{mc} - G_{mc})$ will be practically constant, because this term depends just on the structure of chain and surface. The exclusion is in the case of changing local conformation, as well as the whole chain one.

All the paragraphs did not consider the contribution of conformation variation into the change of process free energy. In general case values of extreme parameters, at which there appears thermodynamic opportunity of reaction initiation, decrease at the formation of thermodynamically instable conformers.

Thus, as it follows from the consideration of this paragraph, thermodynamic opportunity of reaction initiation occurs at the change of conditions. The performance of this reactions is defined by kinetic opportunities of the process. As the conditions of production usually differ sufficiently from these of storage and exploitation of polymers, then in consequence of all above mentioned there occur the opportunities of degradation. The present paper estimated probabilities of degradation initiation for comparatively simple processes. In general case similar approach can be applied to the change of reaction mechanisms, which can prevail in various conditions of storage and exploitation of the sample [27-30].

REFERENCES

1. N.M. Emanuel, A.L. Buchahenko, *Chemical Physics of Polymer Degradation and Stabilization*, VNU Science Press, 1087, 340 pp.
2. G.E. Zaikov, A.L. Iordanskii, V.S. Markin, *Diffusion of Electrolytes in Polymers*, Utrecht, VSP Science Publ., 1988, 322 pp.
3. Yu.A. Slyapnikov, S.G. Kiryushkin, A.P. Mar'in, *Antioxidative Stabilization of Polymers*, Chichester (W. Sussex), Ellis Horwood, 1994, 246 pp.
4. E.F. Vainshtein, *The Role of Intermolecular Interaction in Polymer Degradation Processes*, Polymer Yearbook, Ed. R.A. Pethrick, G.E. Zaikov, London, Gordon and Breach, 1993, vol. 9, p. 69-79.
5. A. Popov, N. Rapoport, G. Zaikov, *Oxidation of Stressed Polymers*, London, Gordon and Breach, 1991, 336 pp.
6. K.S. Minsker, S.V. Kolosov, G.E. Zaikov, *Degradation and Stabilization of Polymers on the Base of Vinylchloride*, Oxford, Pergamon Press, 1988, 526 pp.
7. E.F. Vainshtein, *The Limiting Sizes of the Large Molecules*, Polymer Yearbook, Ed. R.A. Pethrick, G.E. Zaikov, London, Gordon and Breach, 1992, vol. 8, pp. 85-113.
8. Yu.V. Moiseev, G.E. Zaikov, *Chemical Resistance of Polymers in Reactive Media*, New York, Plenum Press, 1987, 586 pp.
9. R.M. Aseeva, G.E. Zaikov, *Combustion of Polymeric Materials*, Munchen, Karl Hanser Verlag, 1986, 390 pp.
10. N. Grassie, G. Scott, *Polymer Degradation and Stabilization*, Cambridge, Cambridge University Press, 1985, 222 pp.
11. E.F. Vainshtein, A.A. Sokolovskii, A.S. Kuzminskii, Kinetics of the Changing Products Molecular-Mass Distribution in

Thermodegradation of Associated Polymers, Polymer Yearbook, Ed. R.A. Pethrick, G.E. Zaikov, London, Gordon and Breach, 1993, vol. 9, p. 79-101.

12. E.F. Vainshtein, G.M. Gambarov, S.G. Entelis, Doklady Akademii Nauk SSSR, 1974, v. 113, N2, p. 375-378.

13. E.F. Vainshtein, Thesis Dr. of Sci., Institute of Chemical Physics USSR Academy of Sciences, Moscow, 1981.

14. L.D. Landau, E.M. Livshits, *Statistical Physics* (in Russian), Moscow, Nauka, 1964.

15. P.J. Flory, *J. Chem. Phys.*, 1942, v. 10, N1, p. 51.

16. H.L. Huggins, Ann. N.-Y. Academy of Sciences, 1942, N43, part 1, p. 1.

17. E.M. Guggenheim, *Mixtures,* Oxford, Clarendon Press, 1952, 272 p.

18. A.J. Staverman, *Rec. Trav. Chim. Phys.-Bas.*, p. 1 (1950).

19. I. Prigogine, *The Molecular Theory of Solution,* Amsterdam, Nort-Holland Publ. Co., 1957, 448 p.

20. A.D. Pomogailo, *Immobilized Polymeric-Metal Complex Catalysts* (In Russian), Moscow, Khimiya, 1986.

21. I. Oshima Takumi, Nagai Toshikazu, *J. Org. Chem.*, 1991, v. 56, N2, p. 673.

22. *Chemical Encyclopedian Dictionary,* Ed. I.L. Knunyants, Moscow. Sovetskaya Encyclopedia, 1983, p. 791.

23. E.F. Vainshtein, G.E. Zaikov, "Chain Effect" in *Processes with Oligomeres, Polymer Yearbook,* Ed. R.A. Pethrick, G.E. Zaikov, London, Gordon and Breachy, 1993, vol. 10, pp. 231-235.

24. E.F. Vainshtein, A.A. Sokolovskii, *Thermodegradation of Associated Polymers,* ibid., 1994, vol. 12, in press.

25. O.F. Shlensky, A.A. Matyukhin, E.F. Vainshtein, *J. Thermal, Anal.* 1988, v. 34, p. 645.

26. A.A. Sokolovskii, E.F. Vainshtein, K.M. Gubeladze, A.S. Kuzminskii, *Vysokomolekulyarnye soedineniya,* 1988, vol. 30B, N1, p. 244-248.

27. G.E. Zaikov, Russian *Chemical Review,* 1993, vol. 62, N6, p. 603-623.

28. G.E. Zaikov, *International J. of Polym. Mater.* 1994, vol. 24, N1-4, p. 1-19.

29. V.V. Kharitonov, B.L. Psikha, G.E. Zaikov, ibid, 1994, vol. 26, N3-4, p. 121-176.

30. G.E. Zaikov, 208th *American Chemical Society National Meeting and Exposition Program,* Washington DC, August 221-25, 1994, ACS Publ. p. 125.

TEMPERATURE EFFECTS ON THERMODESTRUCTIVE PROCESSES

E.F. Vainshtein, G.E. Zaikov and O.F. Shlensky

Abstract— Temperature effect on thermodestructive processes in polymer are considered. The logarithm of life-time of a linear polymer thin film on $1/T$ (inverse temperature) was found to be a smooth increasing curve tending to a limit at definite values of $1/T$, characteristic to every polymer. Temperatures above with avalanche-like decomposition of polymers occurred were called limiting. To explain the dependence it was assumed that associated and non-associated monomer links are destructed with different rate constants the sequence of chemical acts in the process being the same. The rate constant of destruction of non-associated monomer links is higher than that of associated ones. Equilibrium of concentrations of associated and non-associated monomer links is established much quicker than chemical reactions run.

At rather low temperatures the reaction runs presumably at chain ends. Temperature being increased the number (concentration) and the length of blocks consisting from non-associated monomer links increase, either. When the length of such a block reaches a definite value a break reaction occurs, the break constant increasing with the increase of the block length. The dependence of the process rate on temperature is determined both by concentrations of blocks of various length and the temperature dependence of break rate constants. At a definite temperature the reactions presumably at the chain ends are substituted by chance reactions. At temperatures when intermolecular interactions can be neglected thermodynamically unstable "polymer" gas appears which decompose in an avalanche-like manner.

INTRODUCTION

Polymer materials are widely applied in modern technical systems, working under strong heat and power loads. They are affected both by rather low (for a long time) and high temperatures (shock loads for a short time). Because of a wide range of exploitation conditions, physical-chemical, thermodynamical and other properties of materials change. These changes are induced by both chemical processes, destruction, for instance, and changes in aggregate and physical states. To predict the life duration one should evaluate both the temperature effect on chemical and physical processees and their reciprocal influence. The aim of this work is to consider the temperature effect on processes of polymer destruction.

At relatively low temperature dependences of logarithm of thin film lifetime τ on $1/T$ (inverse temperature) are described by the equations close to Arrhenius one [1]. So the main part of experiments was carried out at rather high temperatures. A lot of chemical reactions can run at various temperatures, so the rate of reaching the predetermined temperature should be minimal. Increase temperature dependence of the thin film lifetime was studied.

EXPERIMENTAL METHODICS

Thin films of linear polymer melt were applied onto a metal plate preheated up to a given temperature. The test rod was pressed to the plate surface and moved with constant rate. In some experiments the

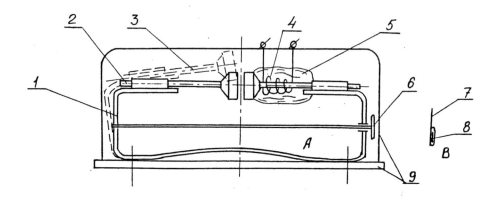

Figure 1. The schematic diagram of the experimental set-up (A). 1 - flat spring; 2 - the heating element in operating position; 3 - the same element in the initial position; 4 - the heating spiral (not shown at the left element); 5 - heat isolation; 6 - stop knot (B); 7 - the package with the sample; 8 - the sample; 9 - the base and protective cover of the heater (removable). The diagram explains the principles of the set-up action. In operating set-up the elements are moved automatically by an electromagnet starter, attained instead of the knot 6.

metal plate was moved relative to the rod. Because of melting a film not more than 7μm thick was formed on the plate surface. The maximal film thickness which permitted its instant heating up to a given temperature was determined experimentally and calculated [2]. In spite of intensive gas evolution especially at high temperature the melt film appeared strongly complied to the plate surface and was not torn even when it was below. Strong contact survived thanks free diffusion of formed gases through the film. The film lifetime on the plate surface was measured either visually or by a thermocouple indications. Its junctions were situated on the plate surface at the point of decomposition. To measure the lifetime the filming could also be used with the following video fixation in IR-light with the help of a video-analysis system, a synchronous pulse generator of SEKAM heat vision monitor [3]. Experimental set-ups for visual or thermocouple watch are shown at Figure 1. A moving specimen used, the most clear picture of the melted film was observed in IR-spectrum as a self-emission of the object was registered. The typical photo is shown at Figure 2.

Figure 2. The photo of the disappearing trace at the plate.

A white spot face follows the specimen and disappears as a comet tail. The length of the tail divided by the movement rate determined the lifetime with high accuracy. At relatively low temperatures (below 570°C) visual and thermocouple measurements were used. The time of the film heating up to the temperature of the plate at relatively high temperatures (below 800°C) did not exceed 0.01s (heat rate $5 \cdot 10^4$ grad/s). Experimental heat time up to a given temperature was stated with the help of heat vision apparatus by watching the film emission intensity changes in the series of video pictures. So, the film thickness being 20µm, the time did not exceed 40ms. The heat time is proportional to the quadrat of the film thickness in accordance with calculations.

DISCUSSION

The results of tests according to the described methodics are exhibited at Figure 3 as dependences $\lg(1/\tau)$ on $1/T$ for a number of linear polymers [3-5]. At low temperatures the dependence is close to a linear one. Lineary of the dependence is characteristic for one-staged (Arrhenius equation) or consecutive processes. At a linear part (below 480°C) the experimental data do not practically deviate from the literature ones [1]. At higher temperatures the dependence deviates from the straight line to a higher rate one. Then the curve goes up sharply tending to a limit asymptotically. Limit temperatures have characteristic values for every polymer. However, polymers of different chemical structures have relatively close limit temperatures. One should note that despite different mechanisms of the processes in different polymers all the dependences observed are qualitatively similar.

As it follows from Figure 3, the process has at least one parallel stage, the rate constant of the one running at higher temperature being higher. From the analysis of the structure and features of reactions in polymer it follows that reactions of associated and non-associated monomer links may appear such parallel processes. Assume the sequence chemical acts in processes close, as these sequences may not change even at changes of aggregate state [6]. The rate of equilibration time between associated and non-associated monomer links is higher than that of chemical reactions because of significant differences in energies of formation of physical (intermolecular) and chemical bonds. With the temperature increase the number of non-associated monomer links obviously increase, the rate constant (K) of the destruction of the type

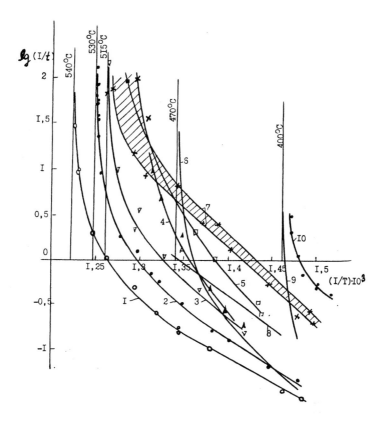

Figure 3. Lifetime logarithm dependence on the inverse temperature for a thin film of a linear polymer.

is higher for non-associated monomer links than for associated ones. The reason is that to remove radicals formed in associated monomer links it is necessary to consume energy to tear off intermolecular bonds: otherwise radical recombination $(\dot{R} + \dot{R} \rightarrow R - R)$ should occur. The observed rate constant for associated monomer links can be close to zero. It means sharp decrease of the break reaction and later practically full absence of it at low temperatures. As the destruction reaction has two rate constants, the rate equation looks as follows:

$$\frac{dM}{dt} = k_1 n_1 N + k_2 n_2 N \tag{1}$$

where

M - The quantity (concentration) of substances isolated in the reaction;

k_1 and k_2 - rate constants of destruction of non-associated and associated monomer links;

N - concentration of macromolecules;

n_1 and n_2 - the number of non-associated and associated monomer links, respectively.

The removal of a substance from the surface is possible only if a low-molecular volatile substance is formed. The break reaction at the end of the chain appears the most favorable, especially at low temperatures. To remove the substance, formed in the middle, two chain breaks are necessary. Note that from thermodynamical point of view the chain end is non-associated to a higher degree. It is because at equal length, the non-associated part at the end of the chain can realize larger number of states than that in the middle. Then at low temperatures, because of increase of association degree, the non-associated monomer links would locate at the chain ends with higher probability. The dimensions of non-associated monomer blocks (the number of non-associated monomer links located in succession and bounded by associated ones) at low temperatures should not be large. Suppose that at low temperatures the reaction should run presumably at the chain ends. In this case more low-molecular products should be formed than at the breaks in the middle of the chain. The products could evaporate quicker than chemical reaction runs. Then the quantity of the volatile substance should be determined by the kinetic equation

$$\frac{dM}{dt} = k_1 N_1 + k_2 N_2 \tag{2}$$

$$N_1 + N_2 = ZN \tag{3}$$

where

Z - the average number of ends in the molecule;

N_1 and N_2 - concentrations of macromolecules with non-associated and associated end groups, respectively.

In all the cases, besides particularly specified, it was supposed that the number of molecule ends in reactions should not change and the destruction should appear uniformly all over the volume.

The equation (3) was derived at the supposition that reaction rate constants at the ends should be close and the degree of destruction should not be too high. At high degree of transformation the rests of chain should evaporate ($N \neq const$). The rate constant of evaporation should obviously increase with the decrease of a substance molecular mass. The rate of evaporation at relatively low temperature is higher than that of chemical reactions.

The ratio of associated and non-associated monomer links is determined in accordance with energy distribution of molecules (by Gibbs)

$$K = \frac{N_1}{N_2} \qquad (4)$$

where

K - thermodynamical equilibrium constant, depending just on temperature.

For equation (4) it was supposed that both equilibrium constant and rate constants should not depend on the location of associated and non-associated monomer links and the chain length and, consequently, on molecular-mass distribution.

One should note that in polymers, especially at low temperatures, this ratio is valid in rare cases because there exist stationary states, depending on production procedure, storage and utilization of the specimen. Under given conditions the specimen state can deviate significantly from an equilibrium one. The temperature increasing, the difference between stationary and equilibrium states of the specimen, produced under certain conditions, should decrease. Meanwhile the ratio of the number (concentration) of non-associated and associated monomer links, located at the ends, should be close to the equilibrium. For end groups the deviation from the ratio of associated and non-associated monomer groups under equilibrium (eq. (4)) should be taken into account by a temperature dependence correction coefficient.

Temperature dependences of k_1 and k_2 are described by Arrhenius equation

$$k_1 = k_1^0 e^{-E_1 RT}; \; k_2 = k_2^0 e^{-E_2/RT}$$

and that for K - by Vant-Hoff equation

$$K = K_0 e^{-H/RT}$$

The solution of the equation set (2-4) with account of temperature dependences of rate and equilibrium rates looks as follows

$$M = ZN \cdot \frac{k_1^0 e^{-E_1/RT} \cdot K_0 e^{-\Delta G/RT} + k_2^0 e^{-E_2/RT}}{1 + K_0 e^{-\Delta H/RT}} \cdot t$$

And inverse temperature (1/T) dependence of $\lg \tau$ (lifetime) is

$$\lg\tau=\lg M-\lg ZN+\lg(1+K_o e^{-\Delta H/RT})-\lg(k_1^o e^{-E_1/RT}\cdot K_o e^{-\Delta H/RT}+k_2^o e^{-E_2/RT}) \tag{5}$$

If the system contains molecules of low-molecular masses, to determine mass losses one should take into account both the molecules themselves and the formed pieces of molecules, which can evaporate.

If one takes into account the above mentioned and the fact that $K\ll1$ (low temperatures), he gets a linear dependence of $\lg\tau$ (lifetime) on inverse temperature ($1/T$), which is observed experimentally [2]. At higher temperatures K becomes comparable with 1, and the observed dependence deviates from a linear one.

At higher temperature there exist more long blocks of non-associated monomer links. Suppose, that the probability of intermolecular bond formation should be $p=n_a/n$, where n_a - the average number of associated bonds. Then the probability that at a given link the bond is not formed, equals $(1-p)$ (the probabilities of intermolecular bond formation at different parts of chain are supposed equal and independent on its length). If equilibrium in the system is established at a given temperature, the probability of existence of the block from m non-associated monomer links (see the specification of the block length) equals:

$$p_m = p^2(1-p)^m \tag{6}$$

The probability of existence of the blocks from t associated monomer links equals

$$p_t = p^t(1-p)^2 \tag{7}$$

The size of the block with associated monomer links is specified as a number of the latters located in succession in the chain and bordered by non-associated monomer links. So, the concentrations of block of different sizes in dependence on their lengths are described by a diminishing geometrical progression. The maximal block length can equal the number of monomer links if one monomer link forms one intermolecular bond. The free energy gain at the break of non-associated blocks is caused by the increase of bending entropy of chain ends formed, the value of entropy change depending on the length of formed ends and initial blocks. If molecules of low-molecular mass are formed (at the break of chain ends) one should take into account just molecule number changes. The bending entropy per a link in the final state is, at least, not higher than in the initial chain part. The maximally attained entropy gain per a bond (monomer link) equals R. In reality it is always lower. So a definite length of a non-associated block is necessary to make the chain break thermodynamically possible. The more the block length, the higher the break rate constant is. The probability of intermolecular bond formation p is connected with the equilibrium constant as follows

$$p = 1/(1 + K) \tag{8}$$

It means that the ratio of associated and non-associated monomer links is determined by the thermodynamics of intermolecular bond formation. Assume all the components of the free energy additive. Then the free energy of chain formation without intermolecular bonds ζ_H equals

$$\zeta_H = \zeta_x + \zeta_{bend} \tag{9}$$

where
ζ_x - free energies of formation of stretched molecules;
ζ_{bend} - the free energy of molecule bending.

Monomer links can exist in various conformations, so one should take into account the free energy of conformer formation that of bending of various conformers and the possibility of their transpositions in the chain. At high temperatures the energy of transition between conformers is not high and to simplify we should not consider this component. The free energy of chain formation ζ_a with n intermolecular bonds equals

$$\zeta_a = \zeta_x + n\zeta_b - RT \cdot \ln W + \sum_i \zeta_{bend_i} \tag{10}$$

where
ζ_b - the free energy of intermolecular bond formation;
W - the number of methods of formation of n intermolecular bonds along the chain, containing n_0 bonds, under supposition that all the bond should be equivalent, ζ_{bend} - the free energy of bending of a block, containing i non-associated monomer links. The block size changes from 1 up to the maximal value n_0.

The bending free energy leads to in equivalence of intermolecular bonds, so it is more valid to evaluate ζ_a according to Gibbs' distribution. This additive consideration of ζ_a induces no deviations from the physical sense of the observed dependence on the film lifetime logarithm.

The minimal number of intermolecular bonds, when the molecule can be associated, got from the condition $\Delta\zeta = \zeta_a - \zeta_n$, equals

$$n_{\min} = (\zeta_{bend} + RT \cdot \ln W - \sum_i \zeta_{bend_i})/\zeta_b$$

The limiting quantities of intermolecular bonds necessary to keep the molecule in the associate [7] or at adsorption [8] are known from literature.

Temperature being increased, ζ_b/RT decreases tending to zero, and ζ_{bend}/RT increases tending to the limit. It leads to $n_{\min}$ increase and the number (concentration) of intermolecular bonds decrease.

In accordance with the energy distribution of molecules the width of the distribution of associated blocks by $n_{\min}$ sizes increase. The maximal size of an associated block is due to the chain length and at a given temperature the molecules do not associate.

The system free energy ζ_{form}, possessing various numbers of intermolecular bonds in chains, is determined by the equation

$$1 = \sum_i p_i = p_0 + p_1 + \ldots + p_{n_0}$$

where

$$p_i = Ae^{-\zeta_i/RT}, \; A = e^{-\zeta_{form}/RT}$$

The change of the free energy in the process, which defines the equilibrium constant, equals

$$\Delta\zeta = \zeta_a - \zeta_n = -RT \cdot \ln K$$

It is obvious that at high temperatures where the dependence of lgK on $1/T$ is not described by a linear anamorphosis of Vant-Hoff equation [9], being a diminishing function, tending to infinitum at the temperature, above which molecules do not associate. Though the main reason for K decrease lies in entropy component of the free energy, enthalpy of the process changes, either, as the average number of bonds per an associate changes.

It is just at high temperatures, when "free rotation" of bonds in the chain appears, that we have

$$\sum_i \zeta_{bend} = (n_0 - n)RT$$

simultaneously the observed local equilibrium constants become equal for different places of chains.

At lower temperatures non-associated monomer links locate at the chain ends with high probability. It means that there exists unequally probable distribution of blocks of a given length along the chain. As every block breaks with its own rate constant, the form and width of the distribution of reaction products can change with temperature change. If at low temperature the reaction runs presumably at the chain ends, at high one the break is equally probable at different places of the chain. It means that at temperature change the process can change its direction, the sequence of chemical acts kept constant (from the reaction at the end groups to the chance one). So, if pyrolysis of polystyrole at relatively low temperatures gives more tan 90% of styrole in the reaction products [10], the destruction of polystyrole with $\overline{M} = 600000$ at temperatures above 530°C gives chain fragments of various molecular masses from dimers to $\overline{M} = 3000$ as the reaction products. It was established with the help of gel-penetrating chromatography. The limit chain length is defined by the possibility of its existence in gaseous phase at given external conditions.

To write down kinetic equations describing the destruction process one should obviously take into account changes with temperature of both reaction abilities of blocks of various lengths, block distribution of non-associated monomer links by sizes and block concentrations. Unfortunately, we don't know such a scheme.

In experiments we considered the mass losses by the whole film. That's why the kinetic description of the process should account the formation of molecules which can evaporate. It makes it necessary to consider the change of concentrations of macromolecules of various lengths. The reaction ability of different places being equal and independent on the chain length, the description of the process should consider not only two rate constants of destruction but rate constants of evaporation of the molecules formed, either. So there exist the following kinds of molecules in the system:
1. The molecules which just decompose with time. The rate of their concentration decrease

$$\frac{dN_n}{dt} = K_1(1-p)nN_n + K_2pnN_n = \tilde{K}nN_n \qquad (11)$$

where N_n - concentration of molecules with n monomer links.
As a rule they are molecules of maximal size

$$\tilde{K} = k_1(1-p) + k_2p$$

Time dependence of concentration for such molecules looks as follows

$$N_n = N_n^0 e^{-n_0 \tilde{K} t} \qquad (12)$$

where N_n^0 - initial concentration of molecules n_0 monomer links long.

2. Molecules, possessing i monomer links ($1<i<n_0$), can not only decompose but be formed from larger molecules (1 - the number of monomer links in the largest volatile molecule). The decomposition rate as in the previous case is obviously proportional to i. Their formation from larger chains a proportional to doubled quantity (concentration) of the latters (the reaction can run with equal probability at both chain ends). The equation for reaction rate looks as follows

$$\frac{dN_i}{dt} = \tilde{K} \cdot i \cdot N_i - 2\tilde{K} \cdot \sum_{j=i+1}^{i=n} N_j \qquad (13)$$

For the simplest case (all the molecules at the beginning of the reaction have equal molecular masses) the solution of the equation is

$$N_i = N_n^0 [(n-i+1)e^{-\tilde{K}it} - 2(n-i)e^{-\tilde{K}(i+1)t} + (n-i-1)e^{-\tilde{K}(i+2)t}] \quad (14)$$

The equation (14) shows that the process is of Markov type, as the number of molecules N_i at a given time moment is affected by their decomposition and formation of new molecules with the number of monomer links (i+1) and (i+2).

If concentrations of the largest molecules just decrease with time, those of intermediates change from zero (t=0), pass a maximum back to zero (t→∞). The time for reaching the maximal N_i value, from the condition $dN_i/dt=0$, equals

$$t_{max} = \frac{1}{\tilde{K}} \cdot \ln \frac{(i+1)(n-i) - i\sqrt{(n-i)^2 + i(i+2)}}{(i+2)(n-i-1)}$$

The maximal concentration value N_i is defined by the substitution $t=t_{max}$ into the equation (14).

On the basis of the law of conservation of matter it was shown [11], that

$$\sum_{j=i+1}^{j=n} N_j = [(n-i)e^{-\tilde{K}(i+1)t} - (n-i-1)e^{-\tilde{K}(i+2)t}]N_n^0 \qquad (15)$$

The time for reaching the maximal value of $\displaystyle\sum_{j=1+1}^{j=n} N_j$ is defined as

$$t_{max} = 1/\tilde{K} \cdot \ln \frac{(i+2)(n-i-1)}{(i+1)(n-i)}$$

Taking into account the principle of independence of chemical reactions, the destruction of macromolecules of a given molecular-mass distribution can be considered as a sum of processes in molecules with definite molecular masses.

3. If molecules consisting from 1 and less monomer links can evaporate, their concentration decrease rate can be written as follows

$$\frac{dN_i}{dt} = (\tilde{K} = k_i) \cdot N_i - 2\tilde{K} \cdot \sum_{j=i+1}^{n} N_j \qquad (16)$$

where
k_i - rate constant of evaporation of molecules with i monomer links ($i \leq 1$). The concentration of evaporating molecules, as any other intermediate particles, changes from zero (t=0) to zero (t→∞) passing through a maximum. This constant depends on temperature by Knutsen-Langmuir law [12]

$$k_i = \frac{k_n \exp(-\Delta G / RT)}{\sqrt{2\pi R / i}}$$

and on molecular mass of evaporating molecules

$$k_i = A\sqrt{i+1} \cdot e^{-b(i+1)}$$

where A and b - parameters of a given homologic row depending just on temperature and heat effect (ΔH).

4. Besides there exist molecules of very small sizes which just evaporate and are formed from larger molecules.

$$\frac{dN_{l_0}}{dt} = k_{l_0} \cdot N_{l_0} - 2\tilde{K} \cdot \sum_{j=l+1}^{j=n} N_j \qquad (17)$$

If molecules can't leave the reaction volume (evaporate), the equations (16,17) don't contain the terms with k_i.

To describe the process rate one should consider changes in concentrations of the molecules of all types, i.e. join the equations (11,13,16,17) in a system. It should consist of a set of linear differential equations with constant coefficients, the number of the equations corresponding to the number of various types of molecules. The solution of the equation system (11,13,16,17) includes the concentrations of molecules of various lengths as a sum of terms, containing exponentials with indexes $\tilde{K}$ and k_j.

The shorter chains, the larger number of terms are included into the solution. Taking into account the accuracy of experimental data during expansion of exponentials in a row we use just initial terms of the row.

At low destruction times ($t \leq 1/k$) the concentration of non-evaporating molecules is

$$N_n = N_n^0 \cdot (1 - n\tilde{K}t), \; N_i = 2\tilde{K}N_n^0 t \tag{18}$$

The mass of the molecule

$$M = \sum_{j=l+1}^{n} N_j \cdot j \cdot M_0 \tag{19}$$

where M_0 - molecular mass of a monomer link.

Mutual solution of equations (18) and (19) and following simplifications give

$$M = M_o'[1 \frac{(l+2)(l+1)}{n} \tilde{K}t]$$

where $M_o' = n_o N_n^o M_o$ - initial mass of the substance.

If $t > 1/k$, the system contains a small number of molecules with high molecular masses. As the processes of exponential character, it is advisable to consider just a term with minimal exponential index. In this case we have

$$N_i = N_n^o(n - i + 1)\exp(-\tilde{K}it) \tag{20}$$

where $l+1 < i < n$.

Substituting (20) and (12) into (19) we have

$$M = M_o' \frac{(n-l)(l+2)}{n+1} \cdot e^{-\tilde{K}(l+1)t} \tag{21}$$

One should note that the solution of equation (21) for volatile components in a linear approximation coincides with that of (18). As a rule, for practical problems with account of experimental accuracy one can consider just the terms up to t^2

$$N_i = N_n^o \tilde{K} t [2 + (n - 2 - 3i) \tilde{K} t - \tilde{K} i t] \qquad (22)$$

The mass of volatile molecules

$$M_{n_i} = M_o' - M_i, M_{n_i} = M_o \cdot \frac{(l+2)(l+1)}{n} \cdot \tilde{K} \cdot t$$

The concentration changes of volatile fragments are defined from the equation

$$\frac{d\tilde{N}_i}{dt} \cong k_i N_i$$

where $i \leq l$.
Hence

$$\tilde{N}_i = \int_0^t k_i N_i dt$$

At low t from the equation (22) we get

$$\tilde{N}_i = N_n^o \tilde{K} k_i t^2$$

The mass of evaporated fragments equals

$$M_n = \sum_{i=0}^{i=l} N_i \cdot i \cdot M_o = N_n^o \tilde{K} M_o t^2 \cdot \sum_{i=0}^{i=l} i A \sqrt{(i+1)} \cdot e^{-b(i+1)} = M_o' / n \cdot K \cdot t^2 \cdot B(l) \qquad (23)$$

where

$$B(l) = A \cdot \sum_{i=0}^{i=l} e^{3/2} e^{-b(i+1)}$$

So, from the above mentioned it follows that the mass of evaporated fragments at the beginning of the process increases weakly because of low amount of small molecules formed (the main number of molecules have the length higher than l). And at the end of the process the number of molecules is small, so the evaporation is low, either.

So, the total time dependence of mass losses is S-shaped, the fact corresponding to experimental data (Figure 4). The changes of molecular-mass distribution of reaction products were evaluated kinetically (by rates of death and formation of the chains of definite length) and statistically.

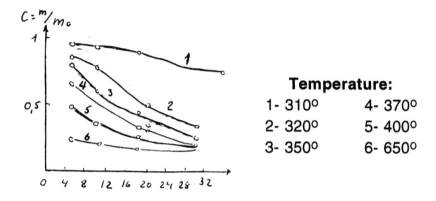

Figure 4. Time dependence of mass losses a thin film of a linear polymer.

When the reaction runs just at the chain ends, for instance, in the simplest case a depolymerization process occurs (destruction at the chain ends), when the largest molecules just decompose

$$\frac{dN_n}{dt} = Z \cdot \tilde{K} \cdot N_n$$

The decomposition is supposed to run at all the chain ends with the same rate constants.

$$N_n = N_n^o e^{-Z\tilde{K}t}$$

The rest of non-evaporating chains, containing i monomer links are formed from the large ones and are ruined during the reaction at the chain ends. The equation for their death is

$$\frac{dN_i}{dt} = Z\tilde{K}N_i - Z\tilde{K}N_{i+1}$$

where $1 < i < m$.

It is supposed that in all the chains the total number of ends does not change. If at the initial moment there existed just the chains of one size, N_n, the change of their concentration with time is described by the equation

$$N_{n-j} = (ZK)^{j-1} \cdot N_n^o \cdot t^j \cdot e^{-Z\tilde{K}t}, \, N_{n-1} = Z\tilde{K}N_n^o \cdot e^{-Z\tilde{K}t}$$

where $2 \leq j < m\text{-}l$.

It is obvious that the time dependence of concentration of molecules of various length (except N_n) has extremum. The maximal value of concentration is at $t = j/Z\tilde{K}$. It equals

$$N_{m-j}^{max} = \frac{N_n^o}{(j-1)!} \cdot j^{j-1} \cdot e^{-j}$$

Maximal concentrations of chains of various lengths decrease with the chain length decrease (just relatively small molecules can evaporate).

Then the equation of their death looks as

$$\frac{dN_l}{dt} = (Z\tilde{K} + K_l)N_l - Z\tilde{K}N_{l+1}$$

where K_l - the rate constant of evaporation.

It is supposed, that molecules formed at the end, evaporate practically immediately. The solution of the equation system, made from the equations of concentration change rates or the molecules of various lengths, can be presented as a sum of terms e^{-ZKt} and $e^{-(ZK+Kl)t}$. The number of terms increases with the decrease of polymerization degree (polycondensation). So, at the beginning the rate of evaporation is practically constant, as the total number of chains does not change. From a definite moment, when in the system there appear volatile molecules of larger size, than those formed in the reaction at the chain ends, the rate of reaction product elimination increases.

The reaction products at the chance destruction are distributed with time by the equation (14). To evaluate molecular-mass distribution by statistic methods we've made the same assumptions as at kinetic consideration [11].

If at the time moment t the average number n' of monomer links is broken, n'$_1$ from them being associated and n'$_2$ - non-associated

$(n'=n'_1+n'_2)$, we can introduce respective probabilities of break $p_1=n'_1/n'$ and $p_2=n'_2/n'$, taking into consideration differences in reaction abilities of associated and non-associated monomer links.

The solution of a system of adjusted equations with account that $n_1+n_2=n$ $(n>>1)$, we have $p_1=n'/n$. The total probability of the break of any bond equals [13]

$$p = p_1 p[H_1] + p_2 p[H_2] \qquad (24)$$

where
H_1 - the event, characterizing the break of an associated monomer link;
H_2 - that for non-associated.

Conditional probabilities of location of associated $p[H_1]$ and non-associated $p[H_2]$ monomer links in the chain are determined as

$$p[H_1] = k/1+k, \ p[H_2] = 1/1+k$$

The total probability in this case equals

$$p = \frac{k}{1+k} \cdot \frac{n'}{n} \cdot \frac{(1+k)\beta}{(1+k\beta)} + \frac{1}{1+k} \cdot \frac{n'}{n} \cdot \frac{(1+k)}{(1+k\beta)} = \frac{n'}{n}$$

Hence, the break probability for any bond in the chain, which has the equilibrium between associated and non-associated monomer links, is the same as in a homogeneous infinitely long chain with non-associated monomer links [14].

From the comparison of changes molecular-mass distribution at destruction by chance and depolymerization, both with account of intermolecular interaction and without it, one can see that weight losses in the case of depolymerization at the beginning of the process are constant, while at chance destruction they increase constantly. If depolymerization is finished, the number of evaporated molecules, larger than monomer, is smaller tan in the case of chance destruction, and equals the number of initial chains. The simplest case of destruction of polymers with the only kind of intermolecular bonds, considered above, can be extended to any number of kinds at chance character of various bond breaks and at chance distribution of these bonds in the chain. If the number of various bonds become equalized quicker than the destruction reaction runs, the probability of break of monomer links with one reaction center will be the same as for the homogeneous infinitely long chain. The equation (24) lets evaluate the probability of bond breaks in inhomogeneous macromolecules, if H_1 and H_2 - events, characterizing the destruction of bonds of various kinds.

If the products remain in the reaction volume, from the equation of formal kinetics

$$-\frac{dt}{dt} = k_1 n_1 + k_2 n_2, \, n_1 + n_2 = n, \, k_1 \neq k_2$$

$$\frac{n_1}{n_2} = k = k_o \cdot \exp(-\Delta H / RT)$$

one can get the break probability for any monomer link in dependence on time

$$p = \frac{n'}{n} = 1 - \exp(\frac{k_1 K + k_2}{1 + K} \cdot t)$$

The temperature of rate constants allows to state the connection of probability with temperature and time. It is difficult to evaluate the input of intermolecular interaction into the destruction basing on time dependence of the probability r another parameter, for example, the number of breaks at constant temperature. So, one should study by temperature dependences.

Let's evaluate the probability of the destruction products of a given length to be located in the system at a given time moment at the fixed degree of transformation. In accordance with Bernulli test scheme by the law of geometric distribution we have

$$p(l) = p(1-p)^{l-1}$$

where p(l) - the probability of occurrence of the fragment with l monomer links, containing (l-1) chemical bond.

The mathematical expectation (M) and dispersion (D) of such a distribution are [13]

$$M = \frac{1-p}{p}, D = \frac{1-p}{p^2}$$

At the same time to evaluate the kinetics of changes of molecular-mass distribution it is better to use the Kuhn distribution [15]

$$p'(l) = p^2 (1-p)^{l-1}$$

where p'(l) - the probability of release of i monomer links in the form of fragments p=n/N.

Note, that the sum of probabilities for the geometric distribution equals 1, and for the Kuhn distribution - p.

Concentrations of monomer links, containing in the fragments of length 1 in both distributions are

$$Cp = Np^2(1-p)^{l-1}$$

The probability of occurrence of fragments of length 1 is maximal at lp=1 and nl=N and equals

$$p_l^{max} = p(1-p)^{\frac{1-p}{p}} = \frac{n'}{n} \cdot \left(\frac{N-n}{N}\right)^{\frac{n-n'}{n}}$$

The time of achievement of maximal probability equal

$$\tau_l^{max} = p(1-p)^{\frac{1-p}{p}} = \frac{n'}{n} \cdot \left(\frac{N-n}{N}\right)^{\frac{n-n'}{n}}$$

The number of broken links being a chance value, described by binomial distribution, the probability for n' bonds in the molecule to be broken is determined by the formula

$$p_n = C_n^{n'} \cdot p^{n'} \cdot (1-p)^{(n-n')}$$

Coefficient   characterizes the possibility of breaking bonds in different places of a chain. Hence, relative mass of fragments containing 1 monomer links is

$$\frac{M_l}{M_0} = \frac{1}{n+1} \cdot \sum_{i=0}^{i=n} C_n^i p^{i+1}(1-p)^{(n-i+l+1)}(i+1) =$$

$$= \frac{1}{n+1}[p(1-p)^{l-1} + p^2(1-p)^{l-2}(n-n'+1)]$$

(25)

As n>>1, the expression (25) is simplified

$$\frac{M_l}{M_0} = \frac{1}{n} \cdot [p(1-p)^{l-1} + p^2(1-p)^{l-2}(n-n')]$$

Mass losses induced by evaporation of fragments of any length from a single monomer link to l_{max} are

$$\frac{M_n}{M_0} = \sum_{i=1}^{i=l_{max}} 1/n[p(1-p)^{i-1} + p^2(1-p)^{i-2}(n-n')] =$$

$$= \frac{1-(1-p)^l(1+pl)}{p^2} \cdot \left[1/n + \frac{n-n'}{n} \cdot \frac{p^2}{1-p} \right]$$

(26)

For low number of breaks (n'<<n) the equation of mass losses looks as follows

$$\frac{M_n}{M_0} = 1 - (1-p)^l(1+pl)$$

The equation (6) allows to determine the nubber of breaks by mass losses and, consequently, molecular-mass distribution of reaction products, if the energy necessary for chain break is supplied so fast that evaporation of the destruction products can be neglected.

At relatively high temperatures (close to limit) during short time periods when destruction does not occur there exist both long blocks from non-associated monomer links and short blocks from associated ones. Evidently, the reactions at the chain ends can be neglected in this case. In large blocks simultaneous breaks of several chain parts are thermodynamically possible. As it was mentioned above, the length of non-associated blocks being increased the rate constant of destruction should increase, either. According to the theory of activated complex the constant increase is due presumably to the increase of the free energy for formation of initial state per a link (the loss of bending free energy). The initial part before the reaction is as if in "excited state." However, the rate constant of destruction can't exceed the value, determined by bond oscillations. As the length of non-associated blocks increase, different parts in them become statistically independent. For such long blocks simultaneous reactions at several places of chain appear possible. Due to it infinite long non-associated blocs (practically, intermolecular interactions in the system can be neglected) avalanche-like polymer decomposition occurs. Mass loss in this case should be defined by the rate of evaporation or carrying away the chain fragments formed.

Similar phenomenon ("spinodal explosion") is rather well known for low-molecular compounds [16]. This explosion is presumably of entropy type, so the rate constant of destruction has a weak temperature dependence. "Spinodal explosion" in polymers is characterized by avalanche-like decomposition of chains (in low-molecular compounds the chemical reaction does not run). That is why polymer molecules similar to low-molecular ones have temperatures above which they can't be heated. Such temperatures widely observed in experiments (Figure 3) were called critical. Evidently, polymers of various chemical structures have various limit temperatures.

In the literature one could fine notes that decomposition of, for example, polyethylene and polytetrafluorethylene should run though explosion due to high negative free energy of the process [17]. However, at relatively low temperatures such character of the process did not occur. It is namely the experimental observation of such explosion at temperatures, where intermolecular interactions can be neglected, that confirms the correctness of interpretation of the role of intermolecular interaction in the destruction.

Local interactions in chains and similar low-molecular compounds being close, the limit temperatures of polymer existence are close to the temperatures of accessible heating of low-molecular compounds. It allows to evaluate them by a modified spinodal equation for Van-der-Waals's medium [18].

To write down the kinetic equations of the process at temperatures close to limit, one should consider the distribution by the lengths of non-associated blocks. Then the equation of destruction rate looks as follows

$$\frac{dW}{dt} = k_1(1-p)ZN + k_2 ZpN + \sum_{i=m}^{i=n_0-2} k_i i p^i (1-p)^2 N + k_{n-1}(n_0-1)^{n_0-1} pN + k_n n_0 p^{n_0} N$$

(27)

The last equation considers the principle of independence of chemical reactions. Two last terms in it describe destruction, if the chain contains one non-associated monomer link or associated monomer links are absent. The low limit of summation m characterizes the possibility of non-associated block break only beginning from a given length. If the reaction at the chain ends can't be neglected, the rate equation of chain end formations looks like

$$\frac{dN}{dt} = \sum_{i=n}^{i=n_0-2} k_i N(1-p)^2 p^i$$

At constant temperature and break constants being independent on block lengths, we have

$$N_t = N_n^0 \exp[k(1-p)^2 \sum_{i=m}^{i=n-2} p^i t]$$

At temperatures equal to limit ones, all the terms in the equation (27) except the last can be neglected. At these temperatures p is close to unity and the rate constant depends weakly on temperature. The time of decomposition for a given degree of transformation equals

$$t = \frac{nKp^{nW}}{1-W}$$

The film exists at the surface only at W values, characterized by the accuracy of the experiment. The time of decomposition becomes equal

$$t = \frac{nK}{1-W}$$

At limit temperatures free energies of formation of associated and non-associated monomer links are equal to

$$D = RT_{\text{lim}} \tag{28}$$

With account of approximations during derivation of the formula (28) T_{lim} has a simple physical sense: at this temperature the free energy of formation of intermolecular bonds (just this case existing in the limit) entropy losses for formation of a single bond are not high. One can consider D the dissociation energy of intermolecular bonds. So, just at temperatures above limit it is possible to consider polymer chains independent. Energies of intermolecular interaction, calculated by experimental data and cited in literature [19], coincide satisfactorily. So, the intermolecular interaction energies for polystyrole and poly-methylmetacrylate, got in this work equal 1.62 Kcal/mole, and from [19] we have 1.625 Kcal/mole.

The above-said is attributed to pyrolysis of high-molecular compounds. The process running in the presence of oxygen or other oxidizing agents, at least one stage of the process is connected with the chain break, despite differences in oxidizing mechanisms (reaction of oxidation not connected with chain break are not considered). If this stage is limiting or comparable by the rate with those determining the rate of the process, one can suppose that the temperature effect on the process should be similar.

Moreover, at the stages with the chain participation one should consider the changes of system entropy, due to changes of mobility.

Rate constants of such stages will depend on temperature in accordance with the changes of chain mobility. These changes lead to the effect similar to "the chain effect", i.e. the rate constant will increase with the increase of the mobility of the reacted chain part [20]. It is clear, that the reaction features without chain break can also be considered according to the above-said.

As the reaction rates at various temperatures are lower than the rates of equilibration, it is impossible to describe the process by the dependence of reaction rate on temperature and therefore on the equilibration rate in the sample without calculation the peculiarities (different rate constants of block reactions, changes in ratios between blocks and their lengths, change of the reaction direction). The attempts to describe the whole kinetics of the processes disregarding the role of intermolecular interactions appeared unsuccessful [21].

It should be noted also that close parts with small non-associated blocks are capable to react with neighbor chains. Radicals, formed in the process, will react with close neighbor chains with higher probability. In associated blocks the reactions with removal of side groups (-H, $-CH_3$, etc.) are possible, either. For more flexible parts the chain break is more probable, as in this case the gain in chain entropy increase and, therefore, the system free energy decreases to higher degree.

Hence, the dependence of the reaction rate on the heat rate is also due to the influence of intermolecular interactions upon the process. If a given temperature is being reached slowly there is enough time for the reactions to occur, which induce changes in chain structure. It is clear that the structure of the sample destructed at a given temperature is defined by the rate of its achievement. Hence, the process depends on the sample thickness. The increase in the sample thickness leads to temperature distribution in it at a given regime of heat energy supply. Due to it in the sample there appears the distribution by the number of intermolecular interactions between chain. At the heat supply increase the number of intermolecular interactions decrease. If this heat is insufficient for nearly full destroy of intermolecular bonds in the sample, a number of chains will have a small amount of such bonds. It means that there can appear molecules or their aggregates where the number of intermolecular bond swill be insufficient for keeping them in the sample. This number increases with temperature increase. So, at a given temperature the dispersion of the sample is initiated. The dispersion temperature is the sample is initiated. The dispersion temperature is always low that the "limit" one. Sizes of dispersed particles depends on the sample mass. In thick samples the particles of presumably given size are formed. The existence of the particles smaller than a given size is thermodynamically disadvantageous [22]. Remind, that the initial stationary states of the samples being identical, their movement to equilibrium and the process of equilibration are simplified with temperature increase. For massive samples even if supplied heat is enough for break of all or nearly all the intermolecular bonds, the rate of temperature equilibration along the sample will induce the initial break of intermolecular bonds at the heat surface and therefore the spaying (dispersion) of the sample. Gaseous products formed at dispersion and destruction and having no time to diffuse to the surface, increase dispersion even more. So, to evaluate "limit temperatures" one should use the films of such thickness that the gaseous products formed should be able to diffuse to the surface. With regard to the above-said, to determine experimentally "limit temperatures" and temperature dependence of the film lifetime the films above 7μm thick were used. It is only for thin films, where thermodynamically stable particles can't be formed, that one can observe the temperature dependence of lifetime due to just chemical reactions. So, the observed lifetime is the function of the sample thickness therefore, to evaluate T_{lim} by calculation and experimentally, one should use the sample of the thickness, corresponding to the above-mentioned demands. This thickness is calculated in [2].

The features of destruction of net polymers are due to absence of chain ends and existence of short segments between knots of branching. The latters being destructed, even shorter segments are formed, which look as "rigid sticks." If the chain ends are absent, the reaction can be thermodynamically probable only beginning from a given temperature. Chain ends, existing in the net, can destruct at lower temperature. It allows to evaluate the concentration and length of such ends. The temperature, where the destruction of chain segments between links can occur, is determined by the equality of the net free energies before and after the break. Tensions in the net are connected with both the distribution of chain segments between links and the number and distribution of chain segments between links and the number and distribution of intermolecular bonds in the sample, because of the lack of flexibility in short chain segments between knots of branching the destruction in these places is similar to the reaction where just low-molecular products participate. For example, the rate constant equals k_0. The chain being more rigid, the rate constant can decrease. Chemical bonds at such segments are more stable. The existence of non-destructing segments in the sample follows from it. Temperature dependence of the destruction degree (Figure 5) is S-shaped [21], as at low temperatures just chain ends destruct, if they exist and at high ones a part of the sample does not destruct due to formation of the system joined from short chain segments. If the segments between knots of branching are too long, after the beginning of the reaction at a given temperature the number of non-destructable parts will decrease with temperature increase, tending to zero at temperatures, close to "limit" ones.

In accordance with the above-said one can observe changes of destruction of joined systems in dependence on net density and segment distribution by length between the knots of branching. Considering the role of flexibility in destruction processes, it is possible to expect that at phase transitions of the second order they should be intensified. In the net of rigid linear polymers the effect is the most pronounced. As this transition (from glass-like to high-elastic state) can limit the reactions from below, the destruction runs relatively intensive just under a given condition (higher temperatures can increase neither the reaction rate, nor the quantity of the precipitate formed). It is possible to explain coking and gasification of coals by the existence of a restricted temperature interval [23].

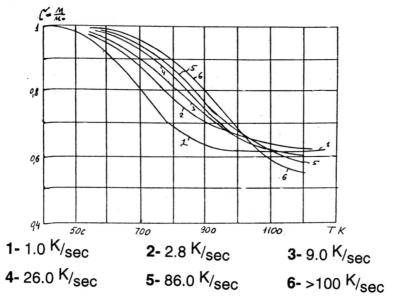

1- 1.0 K/sec   2- 2.8 K/sec   3- 9.0 K/sec

4- 26.0 K/sec   5- 86.0 K/sec   6- >100 K/sec

Figure 5. Time dependence of mass loss for a joined polymer.

Mechanical load can affect the destruction in various ways [24]. As the lifetime of non-associated fragments is small, if the mechanical load is fed during a very short time or is stored beforehand, the temperature of explosive (avalanche-like) destruction should decrease significantly. If the energy is consumed just to break intermolecular bonds, the temperature of explosive destruction is defined by the equation

$$A + RT = RT_{\text{lim}}$$

N.S. Enikolopov et al [25] have observed the sharp increase of destruction rate constant at mechanical load application. It should be noted that temperature affects both k_0 and the chain flexibility [26]. If the mechanical load is constant, the energy is consumed to change flexibility nd to break intermolecular bonds. At low loads the main changes touch flexibility and, hence, the number of intermolecular bonds increase. According to the above-said it can induce the increase of the observed rate of the process [27]. The increase of the load may at first induce deformation of valent angles and then the bond lengths, either. It increases the observed process rate [28]. All above-said concerns both organic and non-organic polymers.

REFERENCES

1. O.F. Shlensky, A.A. Matyukhin, E.F. Vainshtein, *J. Thermal. Anal.*, V. 31, 1986, p. 107.
2. O.F. Shlensky, E.F. Vainshtein, A.A. Matyukhin, *J. Thermal. Anal.*, V. 34, 1988, p. 645.
3. D.N. Yunoev, A.A. Lash, E.F. Vainshtein, et al. *Teplofizika vysokihk temperature*, V. 27, N 2, 1988, p. 369.
4. O.F. Shlensky, E.F. Vainshtein, Dokl. Akad. nauk SSSR, V. 281, N3, 1984, p. 164.
5. E.F. Vainshtein, O.F. Shlensky, *Plastmassy*, N4, 1986, p. 45.
6. N.M. Emanuel, G.E. Zaikov, Z.K. Maizus, *Oxidation of Organic Compounds. Medium Effect in Radical Reactions*, Oxford, Pergamon Press, 1984, 612 p.
7. V.A. Kabanov, I.M. Papisov, Vysokomol. soed. VA21; N2, 1979, p. 243.
8. V.M. Komarov. Thesis, Moscow, Inst. of Chem. Phys., 1982, 28 p.
9. T.B. Varshavskaya, E.F. Vainshtein, *Termodinamika organicheskikh soedinenii (Thermodynamics of Organic Compounds)*, Gor'kii, Gor'kii University Publ. 1989, 189 p. (in Russian).
10. R.M. Aseeva, G.E. Zaikov, *Combustion of Polymer Materials*, München, Karl Hanser Verlag, 1985, 390 p.
11. E.A. Baranova, E.F. Vainshtein, A.A. Matyukhin in: *Modulation and Automatization of Technological Processes in Agriculture* (Modelirovanie i avtomatizatsiya technologicheskikh protsessov v cel'som khozyaistve), ed. E.F. Vainshtein, Moscow, Gosagroprom, 1987 p. 10 (in Russian).
12. Yu.V. Polezhaev, F.B. Yurevich, *Teplovaya zashchita (Thermal Protection)* Moscow, Energiya, 1976, 392 p. (in Russian).
13. I.N. Bronshtein, K.A. Senendyaev, *Spravochnik po matematike (Handbook in Mathematiks)*, 4th ed., Moscow, Nauka, 1986, 544 p.
14. W. Kuhn, Ber. Bd. 63; 1930, S. 1503.

15. N. Grassie, *Khimiya protsessov destruktsii polimerov* (*Chemistry of Polymrs Degradation*) Moscow, IL, 1959, 251 p. (in Russian).
16. V.N. Skripov, B.N. Sinitsyn, P.A. Pavlov et al. *Teplofizicheskie svoistva zhidkostei v metastabil'nom sostoyanii* (*Thermophysical Properties of Liquid in Mtha-Stable State*), Moscow, Khimiya, 1980, 287 p. (in Russian).
17. H. Savada, *Thermodinamika polimerizatsii* (*Thermodynamics of Polymerization*), Moscow, Khimiya, 1979, p. 312.
18. E.F. Vainshtein, O.F. Shlensky, N.N. Lyashikova, Inzhenerno-Fiz. Zh. V 53, N5, 1987, p. 774.
19. A.A. Ascadsky, *Structure, svoistva teplostoikikh polimerov* (*Structure and Properties of Thermostable Polymers*), Moscow, Khimiya 1981, p. 320 (in Russian).
20. E.F. Vainshtein, Thesis, Moscow, Ins. of Chem. Phys., 1981, p. 317.
21. O.F. Shlensky, L.N. Aksenov, A.G. Shashkov, *Teplophizika razlogayushchikhsya materialov* (*Thermophysics of Degradated Materials*), Moscow, Khimiya, 1980, 187 p. (in Russian).
22. L.D. Landay, E.M. Livshits, *Statisticheskaya sizika* (*Statistical Physics*), Moscow, Nauka, 1964, 560 p. (in Russian).
23. M.G. Skylar, *Intensifikatsiya koksovaniya i kachestvo koksov*, (*Intensification of Coke Formation and Quality of Cokes*), Moscow, Metallurgiya, 1976, 997 p. (in Russian).
24. A.A. Popov, N.Ya. Rapoport, G.E. Zaikov, *Okislenie orientirovannykh n napryazhennykh polimerov* (*Oxidation of Stressed Polymers*), Moscow, Khimiya, 1987, p. 232.
25. N.S. Enikolopov, N.A. Fridman, Dokl. Akad. nauk SSSR V. 290; N1, 1986, p. 99.
26. A.A. Popov, N.Ya. Rapoport, G.E. Zaikov, *Effect of Stress on Oxidation of Polymers*, New York, Gordon and Breach, 1990, 360 p.
27. N.M. Emanuel, G.E. Zaikov, V.A. Kritsman, *Tsepnye reaktsii. Istoricheskii aspect.* (*Chain Reactions, Historical Aspect*), Moscow, Nauka, 1989, 336 p. (in Russian).
28. A.A. Popov, Thesis Dr. of Sci., Moscow, Inst. of Chem. Phys., 1987, 320 p.
29. G.E. Zaikov, *International Journal of Polymeric Materials*, v. 24, N 7, 1994, p. 1-38.

THE THERMO-OXIDATIVE DEGRADATION OF POLYSTYRENE

I.C. McNeill

Polymer Research Group, Chemistry Department,
University of Glasgow, Glasgow G12 8QQ, UK

Abstract— The kinetics of thermal and thermo-oxidative degradation of Polystyrene (PS) in presence of 4,4 Isopropylidene bis (2,6-dibromophenol) [DBP] was investigated. It was shown that retardant affects the thermo-oxidative degradation of PS. Thermal volatilization analysis (TVA), subambient TVA (SATVA), FTIR IR spectroscopy, GC-MS technique were used for detailed analysis and determination of the degradation products. The identified products of thermo-oxidative degradation were CO_2, H_2O, styrene, benzaldehyde, methylstyrene, phenol, benzenacetaldehyde, acetohenone, etc.

INTRODUCTION

The general mechanism of thermal and photo-oxidation of PS has been extensively studied [1-6]. As compared with these types of degradation there has been insufficient investigation thermal-oxidative degradation to give a generally accepted mechanism for the reaction.

For example, J. Mullens et al [7] studied the product of reaction in the temperature region about 200°C only. V. Gol'dberg et al [8] and Iring et al [9] studied this process at the same temperature at oxygen pressures from 3.3 to 100 kPa. They suggested mechanism of degradation including the accumulation of hydroperoxide in the first step of reaction. But when a temperature about 300°C or above the presence of this compound did not observe. May be the concentration of the stable hydroperoxide is very low.

The aim of the present paper is an attempt to resolve this problem, to suggest mechanism of degradation and to investigate the influence the retardant in this process.

EXPERIMENTAL

Materials

The second type of PS (PS2) was prepared by free radical polymerization. Polymerization of styrene was carried out at 65°C in vacuum sealed dilatometer in the presence of 0.56% w/v a - Azoisobutyronitrile (AIBN) initiator (a thermostat tank controlled to + - 0,1°C). After approximate 8 h., the dalatometer was removed from the thermostat tank and cooled. The polymer was precipitated in methanol and filtered. It was redissolved in THF and precipitated again in methanol. PS was dried under vacuum at 40°C.

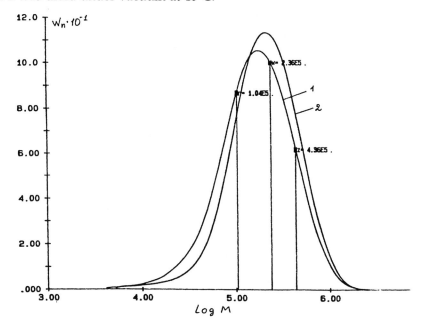

Figure 1. Molecular weight distribution samples of PS1 (1) and PS2(2).

Gel permeation chromotographic (GPC) measurements indicated, that there is a clear difference between the molecular mass distributions of these samples. The polymer characteristics are listed in Table 1 (Figure 1).

Table 1.

PS	Mw	Mn	Polydispersity
PS1	23.000	103500	2.25
PS2	270000	131000	2.06

The free radical polymerization initiator AIBN (Aldrich) was purified by recrystalising twice from absolute methanol, the solution being filtered hot to removed insoluble products of decomposition of the initiator. The crystals was filtered off, dried under vacuum and kept in the dark at 0°C.

It is necessary to mark that THF (May & Baker) did not contain peroxide. 4,4-Isopropylidenbis (2,6 dibromophenol) [Aldrich] was used in work too.

INSTRUMENTATION

Infrared spectra were recorded on a spectrometer Nicolet FTIR Magna-IR 550. Thin films of the polymers (~20mm) wee cast onto aluminum plates (d-25mm) and were used directly for observation of kinetic of process. Reflectance spectroscopy was applied to analyze materials - FT-30 Specular Reflectance Accessory fixed at a 30° angle. Pellets of polymer into KBr were used too, but only for the identification of the products of the degradation PS and IDBP.

The thermal degradation studies were carried out using programmed heating at 5,10,20°/min to 500°C and isothermal heating at 390°C. TVA was carried out under vacuum as described in the paper [10]. An advantage of TVA is that it is non-destructive of various product fractions, which are available for further study, e.g. by spectroscopic analysis. The polymers were examined as film on aluminum plate (weight samples 50-80 mg) and IDBP as powder.

Thermal oxidative degradation was also carried out in TVA apparatus in an air, using a programmed heating rate of 10°/min.

The kinetic of thermal oxidative degradation of PS films and IDBP powder were investigated into Neytech 25 PAF (V=3,5,1) with air exchange furnace at (300+5)°C.

Non condensable gas was analyzed by using a mass spectrometer (VG Micromass QX 200) on line with the TVA system. Condensable volatile degradation products were collected at -196°C with subsequent separation of the volatile products by SATVA. The cold ring fraction (CRF), liquid fraction (LF) and residues were separated and analyzed using GC-MS.

The weight losses of samples was checked scale Unimatic CL-41.

RESULTS

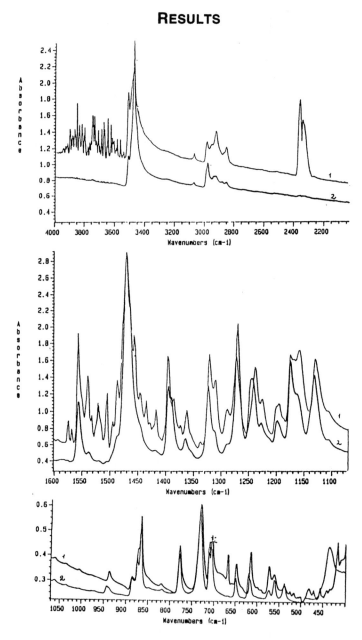

Figure 2. IR-spectrum of IDBP before (2) and after (1) thermal oxidative degradation (T=300°, 2h).

Thermal degradation

The main features of thermal degradation of PS has been established and discussed in the previous paper [2]. The aim of the present experiments were the comparison of the influence IDBP on this process. Before the investigation of polymer + IDBP (9,5% w/w) system, thermal degradation of the pure retardant was studied. SATVA analysis did not show degradation of IDBP in both thermal and thermal oxidative conditions, only sublimation of original retardant is observed. IR-spectra IDBP before and after sublimation are shown in Figure 2. In the next Figure (Figure 3) the typical SATVA curve of IDBP is represented. IR analysis indicates that in the gas and liquid fraction are present only CO_2 and H_2O. After this investigation we move on results of various programmed heating TVA and degradation under isothermal conditions of samples PS. The character of curves (Figure 4) clearly indicate that decomposition of PS and PS+IDBP proceeds with similar behavior and have a single stage degradation. In all cases the 0 and -45 traces are separated from the rest of the traces (-75, -100 and -196°C) which remain with the base line. This indicates that there is no formation of noncondensable products and very volatile condensable products. These results agree with what we had already [2].

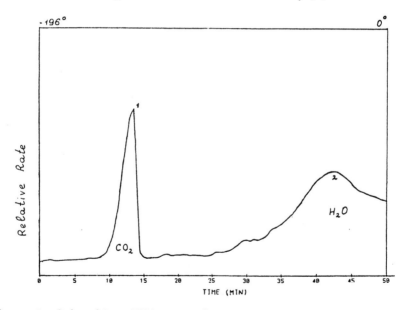

Figure 3. Subambient TVA curve for warmup from -196° to 0° of condensable volatile degradation products from IDBP (thermal degradation programmed heating 25->500° at 10° C/min under TVA condition).

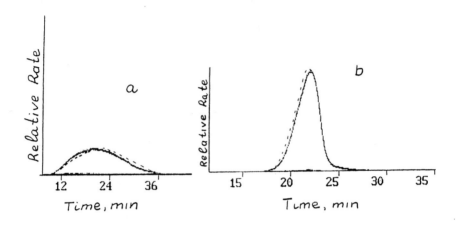

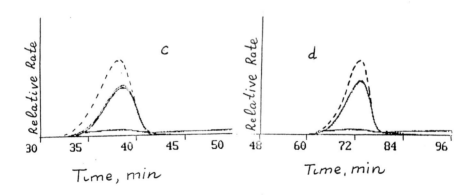

Figure 4.   TVA behavior of PS2 (___) and PS2+IDBP (----) under isothermal heating at T=390° (a) and programmed heating at 20(b), 10 (c) and 5(d)° C/min. Only the 0° Pyrani is shown.

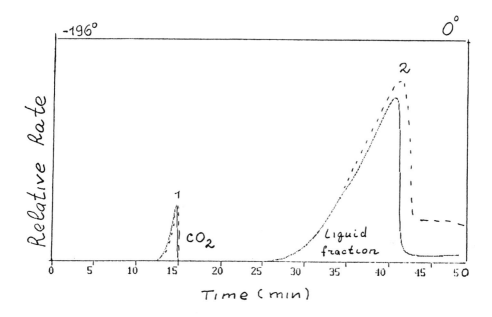

Figure 5. Subambient TVA curve for warm up from -196°C to 0°C from condensable volatile degradation products from PS2+IDBP (thermal oxidative degradation T=300°, 2h.).

Thermal oxidative degradation

The analysis of residue and decomposition products have been made using SATVA, GC-MS and IR-spectroscopy. The SATVA curves illustrated in Figure 5 consist of two peaks. The first peaks (for PS and PS+IDBP) were found by IR-spectroscopy and MS to be carbon dioxide. The second peaks were due to liquid products were identified by GC-MS and contained a mixture of products (Figure 6) of which the main component are represented in Table 2. For example in Figures 7-8 are shown MS-spectra a few identified compounds.

The residue consisted of black insoluble material, about 20% weight. The analysis of residue and the sublimination fraction did not give a definite answer about the products. It can only be said that IDBP is in the sublimation fraction.

As compared with thermal degradation during thermal oxidative degradation there are different rates of process for samples with and without IDBP. In the second case weight losses are faster. Plots of weight losses versus time are shown in Figure 9.

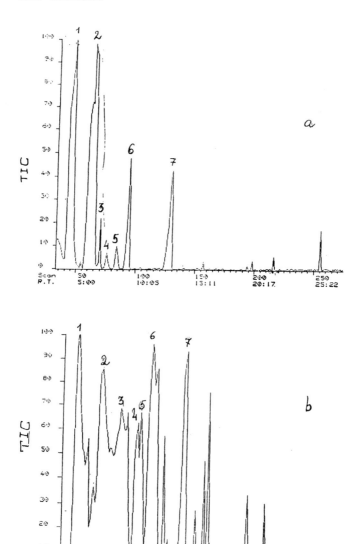

Figure 6. GC chromatogram in GC-MS investigation of the liquid fraction of volatile products of degradation of PS2 (a) and PS2+IDBP (b) under thermal oxidative degradation (air, T=300°, 2h); 1 - styrene, 2 - benzaldehyde, 3 -α-methylstyrene, 4-phenol, 5-benzacetaldehyde, 6-acetophenone, 7-2-phenylpropenal.

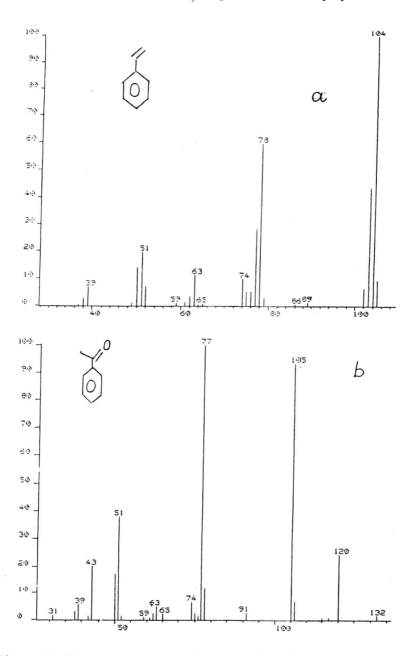

Figure 7. Mass spectrograms of separated products from thermal oxidative degradation of PS; styrene (a) and acetophenone (b).

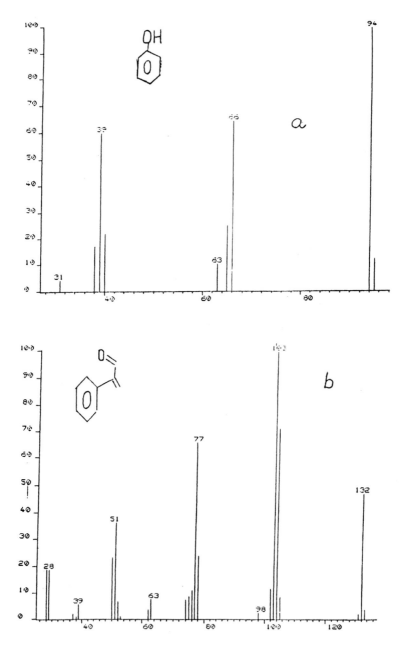

Figure 8. Mass spectograms of separated products from thermal oxidative degradation of PS; phenol (a), 2 Phenylpropenal (b).

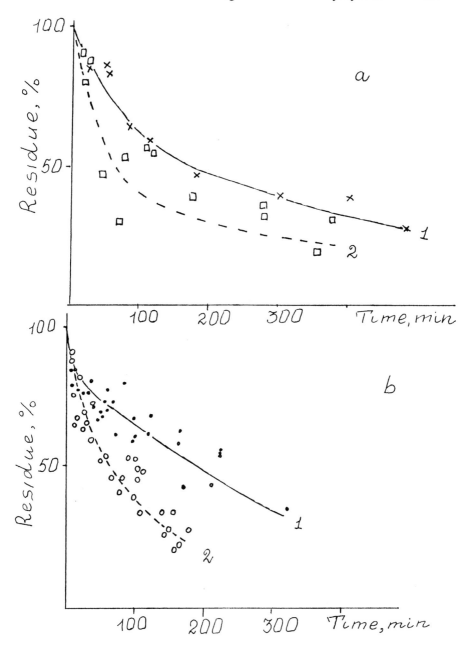

Figure 9. Kinetics weight losses of PS1 (a) and PS2 (b) with (1) and without (2) IDBP.

The analysis of results show that if the residue weight is subtracted from the polymer weight in moment t, it is possible to describe the weight changes during thermal oxidative degradating using the first order rate equation. In Figure 10 the experimental values are shown in the coordinate axes of first order rate.

This result agrees with kinetic degradation of PS composition with hexabromdiphenyl at 295° in air [11]. Same as authors this and other papers [8,9] we think that diffusion and concentration of the oxygen must influence the rate of this process, therefore it is necessary to carry out investigations in the large volume with strong ventilation (into the intra diffusion region). If the ventilation was restricted (the plate with samples was covered with the glass sinter fannel) the rate of degradation decreases.

In Figure 9 can be seen that the degradation rate of PS with retardant for both types PS is slower than pure PS.

Infrared spectroscopic study

The IR spectra original PS, retardant and PS+IDBP after 2h. degradation are represented on the Figures 11-12. As compared with spectra PS and IDBP , the spectra PS+IDBP after degradation have a few new characteristic absorption bands, for example 1684, 1775, 1259 cm^{-1} etc., but in this spectrum the bands at 3470, 1495, 1450, 1475 and 1375 cm^{-1} are weak or absent. The characteristic absorption bands at 3470 cm^{-1} (OH-vibration) and 1475 cm^{-1} (OH or Br vibration) can be connected with disappearance of IDBP in polymeric matrix. The IR-spectra of the band at 3470 cm^{-1} in different moments in time are shown in Figure 13 and the change of concentration IDBP calculated from experimental values illustrates Figure 14. The absorption bands 3470 and 1475 cm^{-1} decrease similarly and after 20 min only less 10% of the original IDBP remains in the plate.

Other results were obtained when bands at 1495 and 1450 cm^{-1} (ring stretch C-C) were analyzed. The intensity of these bands decrease slowly in comparison with the first pair and their change same as weight losses (Figure 15). The rate of change of intensity band at 1375 cm^{-1} (CH or CH$_2$ - vibration) is between first and second pair of bands and describes the equation for the first order reaction.

The constant rates of reaction (k) has been calculated using this band were for PS2 and PS2+IDBP correspondently 4.4 10^{-4} and 2.3 10^{-4} s^{-1}. This values of k are higher than k obtained using bands at 1495 and 1450 cm^{-1} (2.4 10^{-4} and 0.9 10^{-4} s^{-1}). It was not surprise because all authors note that α-hydrogen atom light subjects the influence of the oxygen attack than hydrogen atoms in the aromatic ring. In the presence of IDBP the disappearance of the band at 1375 cm^{-1} proceed slower (Figure 16).

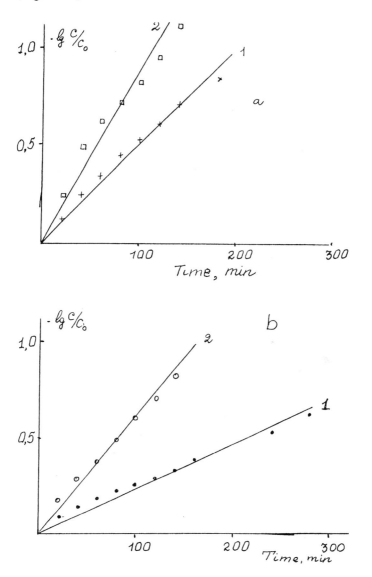

Figure 10. The experimental results in coordinates the first order reaction: PS1 (a) and PS2 (b); with (1) and without (2) IDBP.

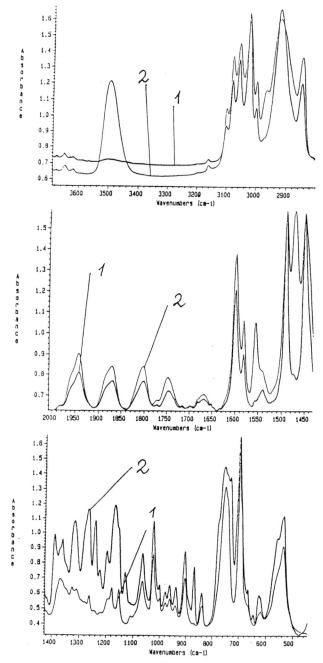

Figure 11. IR-spectrum original PS2 (1) and IDBP (2).

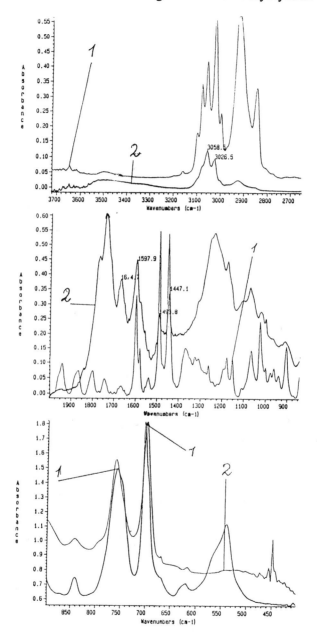

Figure 12. IR-spectrum original PS(1) an dPS+IDBP (2) after thermal oxidative degradation (T=300°, 2h.).

Intensity other bands characteristic for carbon-hydrogen vibration in main chain (2920 and 2830 cm^{-1}) decrease too with transition from original material to residue (Figure 12).

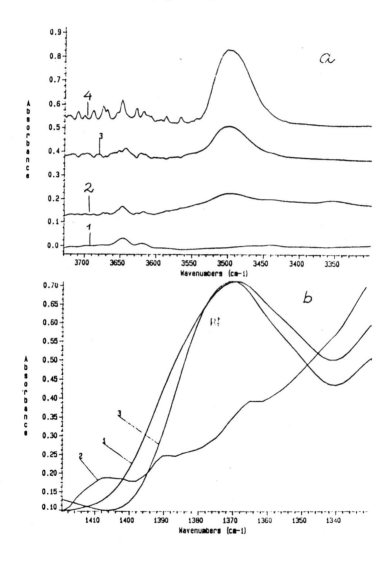

Figure 13. IR-spectrum bands at 3470 (a) and 1370 (b) cm^{-1}. a) original PS2 (1) and PS2+IDBP before (4) and after degradation (T=300°, 2h.) - 2 h. (2) and 10 min (3).

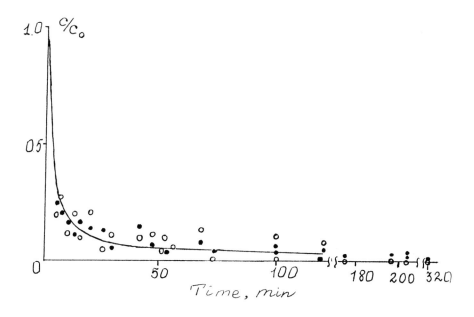

Figure 14. Kinetic curve disappearance of intensity bands at 1475 (o) and 3479 (•) cm^{-1}.

Together with disappearance this bands in IR spectra RS and PS+IDBP observes new band at 1684 cm^{-1}. It is interest to mark that the kinetic curve for this band has maximum (Figure 17). Another special feature of this band as compared with other new bands (1775 and 1259 cm^{-1}) consists in the absence of the visible induction period (Figure 19). Usually the band at 1684 cm^{-1} is assigned to acetophenone structure and the bands at 1775 and 1259 cm^{-1} correspondently to keto-lactone ring (C=0 vibration) and ester group (C-O-C). The analysis of this peak are very difficult because there are the interference several bands by each wave number (Figure 20). The accumulation of peaks at 1775 and 1259 cm^{-1} in the different moment time are shown in Figure 21. Both bands have similar behavior: 1) by ~100 min the intensity are close to constant; 2) the dependence of intensity from time has induction period. The existence of induction period means that the change of intensity of bands at 1259 and 1775 cm^{-1} connected with secondary products of the reaction. In Figure 21 it can be seen that limiting intensity of bands at 1259 and 1775 cm^{-1} are similarly for PS and PS+IDBP. In the earlier

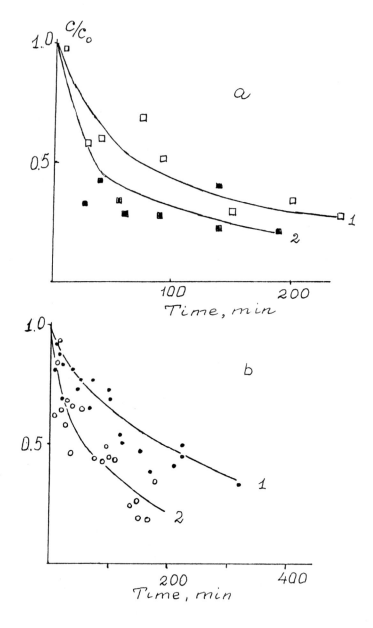

Figure 15. Kinetics of decrease of intensity bands at 1495 and 1450 cm⁻¹ (points) PS1 (a) and PS2 (b) with (1) and without (2) IDBP. Curves experimental results obtained from the weight losses.

stage of process the intensity of band at 1775 cm^{-1} increased faster for PS+IDBP than original PS. Several authors think that in this region absorb keto-lactone type compound. We also suggest that band at 1259 cm^{-1} connected with C-O-C groups, but band at 1775 cm^{-1} - with acid groups [12]. The ring opening mechanism accompanying formation by high temperature stabile OH-group is not possible. Besides, together with accumulation band at 1775 cm^{-1} we observed disappearances OH-groups (Figure 13).

Discussion of Results

The volatile products composition of PS consists from important information on the mechanism of this process. The simplest variant of oxidation is realized under large concentration of oxygen when alkyl radicals (R) rapidly conversion the peroxy ones (RO$_2$). Their participation in reaction of termination and propagation of kinetic chains are quite small. The qualitatively expression of this issue is described as

$$\frac{K_1[O_2]}{\sqrt{k_4}} >> \frac{K_2[RH]}{\sqrt{k_6}} \tag{1}$$

k$_1$ - constant of rate of the interaction reaction between O$_2$ and R˙
    resulted in the yield of RO$_2$
k$_4$ - constant of rate of the quadratic recombination R.
k$_2$ - constant of rate of the propagation of the kinetic chain by radical

$$R\dot{O}_2 + RH \rightarrow ROOH + R\cdot$$

k$_6$ - constant of rate of the quadratic recombination RO$_2$

$$R\dot{O}_2 + R\dot{O}_2 \rightarrow ROH + (R)_2C =) + O_2$$

RH - concentration of active C-H bonds in polymer

The rate of reaction of the oxidation would have a zero order by the oxygen concentration according (1), all products of the radical transformations should have an oxygen-containing groups.

It should be pointed out that under temperatures below 100°C and the concentration of oxygen dissolved in polymer close to the partial pressure in atmosphere (~1.10^{-3} mol O$_2$/kg of polymer under 150 mm Hg), the eq. (1) is correct for the saturated polymers. However, the right part of (1) is arising. Thus, in the thermal oxidation of PS at 180-210°C, the oxygen pressure following which the zero-order oxidation rate for oxygen started increased from 100 to 200 mm Hg. Under the conditions analyzed, during air oxidation at 300°C, oxygen is unlikely to be sufficient to convert all radicals R˙ to RO₂˙.

Furthermore, comparison with the data available in the literature, for air oxidation of polystyrene, the thickness of a film (l=20 mk) will cause a restraint on the reaction rate - the "kinetic" thickness at 200°C is l<10 mk. Due to a resulting decrease in concentration of oxygen across the whole width of a layer, the radicals R˙ shall be ever increasing in importance for formation of oxidation products and thermal degradation [13,14].

This agrees well with the composition of volatile PS destruction products with or without the additive. In either case, significant quantities of hydrocarbons, apparent products of radical R˙ conversion. The likely way of their formation is as follows: the detachment of an hydrogen atom from PS can give rise to 2 types of median alkyl radicals:

a)
$$R - CH_2 - CH - \overset{\centerdot}{C}H - CH - R$$
$$\qquad\quad | \qquad\quad |$$
$$\qquad\quad Ph \qquad\quad Ph$$

b)
$$R - CH_2 - \overset{\centerdot}{C} - CH_2 - CH - R$$
$$\qquad\quad | \qquad\qquad |$$
$$\qquad\quad Ph \qquad\qquad Ph$$

In isomerization with break, the radical "a" gives

$$R - CH_2 - \overset{\centerdot}{C}H \text{ and } CH_2 = C - R$$
$$\qquad\quad | \qquad\qquad\quad |$$
$$\qquad\quad Ph \qquad\qquad\quad Ph$$
$$\qquad (1) \qquad\qquad\qquad (2)$$

from radical "b" gives rise

$$R - CH = CH \qquad \overset{\centerdot}{C}H_2 - CH - R$$
$$\qquad\quad | \qquad\qquad\qquad |$$
$$\qquad\quad Ph \qquad\qquad\qquad Ph$$
$$\qquad (3) \qquad\qquad\qquad (4)$$

Radicals 1 and 4 are terminal macroradicals whose depolymerization yields styrene.

Isomerization of radical "a" with break when valence is near the end of the macromolecule is one of possible ways of α-methylstyrene appearance.

$$CH_3 - CH - \overset{\cdot}{C}H - CH - CH_2 - R \rightarrow CH_3 - C = CH_2 + \overset{\cdot}{C}H - CH_2 - R$$

$$\underset{Ph}{\overset{I}{}} \quad \underset{Ph}{\overset{I}{}} \qquad \underset{Ph}{\overset{I}{}} \quad \underset{Ph}{\overset{I}{}}$$

Other hydrocarbons result from more complex reactions connected with the removal of phenyl groups from chain.

The major oxygen-containing products in the liquid fraction are benzaldehyde, acetophenone, 2-phenylpropenal, benzacetaldehyde, etc.

Benzacetaldehyde and acetophenone appear to be formed after depolymerization of their derivatives due to the formation of radicals from "a" and "b", their conversion to the corresponding hydroperoxides, the appearance of alkyl radicals as a result of their decomposition by the reactions:

$$\overset{\overset{\displaystyle \cdot O}{\displaystyle I}}{R - CH_2 - CH - CH - R} \rightarrow R - CH_2 - HC - C = 0 + \overset{\cdot}{C}H - R$$

$$\underset{Ph}{\overset{I}{}} \qquad\qquad \underset{H}{\overset{I}{}} \quad \underset{Ph}{\overset{I}{}}$$

$$\overset{\overset{\displaystyle \cdot O}{\displaystyle I}}{R - CH_2 - C - CH_2 - CH - R} \rightarrow R - CH_2 - C = 0 + CH_2 - CH - R$$

$$\underset{Ph}{\overset{I}{}} \qquad\qquad\qquad \underset{Ph}{\overset{I}{}}$$

Benzaldehyde results from radical 1 conversation

$$R-CH_2-HC-\overset{\overset{\displaystyle H}{\displaystyle I}}{O} \rightarrow R-CH_2+\overset{\overset{\displaystyle I}{\displaystyle}}{C}=0$$
$$\underset{Ph}{I} \qquad\qquad \underset{Ph}{I}$$

The likely way of 2-phenylpropenal is the decomposition of the radical from benzacetaldehyde derivative

$$R-\underset{\underset{Ph}{I}}{CH}-CH_2-\underset{\underset{Ph}{I}}{\dot{C}}-C=0 \rightarrow R-\underset{\underset{Ph}{I}}{\dot{C}H}+CH_2=\underset{\underset{Ph}{I}\;\underset{H}{I}}{C}-C=0$$

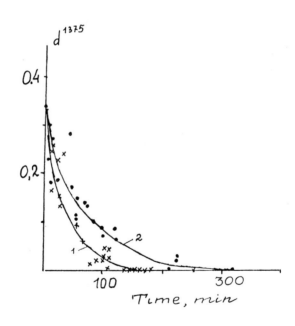

Figure 16. Kinetics disappearance of intensity band at 1375 cm^{-1}; PS2 (1) and PS2+IDBP (2).

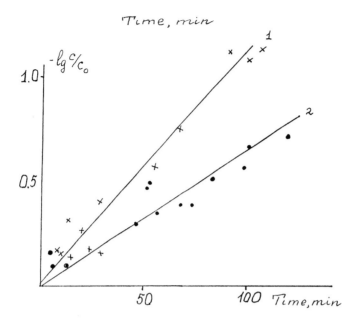

Figure 17. Experimental results disappearance of intensity bands at 1375 cm^{-1} in coordinates of first order reaction: PS2 (1) and PS2+IDBP (2).

The decomposition of another radical yields benzaldehyde itself.

$$R - CH - \dot{C}H - CH - C = 0 \rightarrow R - C = \dot{C}H + CH_2 - C = 0$$
$$\begin{array}{cccccc} I & I & I & I & I & I \\ Ph & Ph & H & Ph & Ph & H \end{array}$$

The acetophenone derivative evolves acetophenone through the following reaction:

$$R - \dot{C}H - Ch - CH_2 - C = 0 \rightarrow R - CH = \dot{C} + CH_3 - C = 0$$
$$\begin{array}{ccc} I & I & I \\ Ph & Ph & Ph \end{array}$$

Our attention is engaged by such an untrivial fact as a relative increase in the proportion of volatile oxygen-containing products of degradation in the presence of the additive. Thus, at regards to the styrene peak which is approximately equal in both cases, the amounts of acetophenone and 2-phenylpropenal are much higher despite the fact that the addition inhibits degradation as shown in Figures 9, 15, 16 and 20.

Here, specific features of diffuse oxidation are likely to manifest themselves. Without addition, oxidation (oxygen uptake) occurs more rapidly and it is concentrated in a relatively narrow area near the surface, wherein it proceeds at a rather high rate up to deep oxidation of CO_2 and H_2O (Figure 6, 5).

In additive-inhibited oxidation, oxygen extends deep into the layer, yet its concentration is lower here and oxidation does not attain high levels. Thus, the effect observed is due to the decreased concentration of oxygen, but its extension into great depths.

It should be noted that the difference in the reaction rates of two PS samples is likely to be also associated with oxygen diffusion into a polymer matrix. The greater the molecular weight of PS is, the lower diffusion coefficient of oxygen, the smaller the reaction layer and the higher oxygen concentration in it.

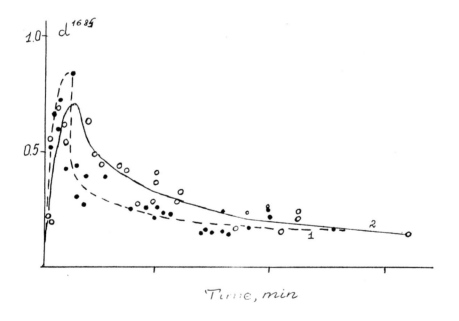

Figure 18. The experimental results, changing intensity band at 1685 cm^{-1}; PS2(1) and PS2+IDBP (2).

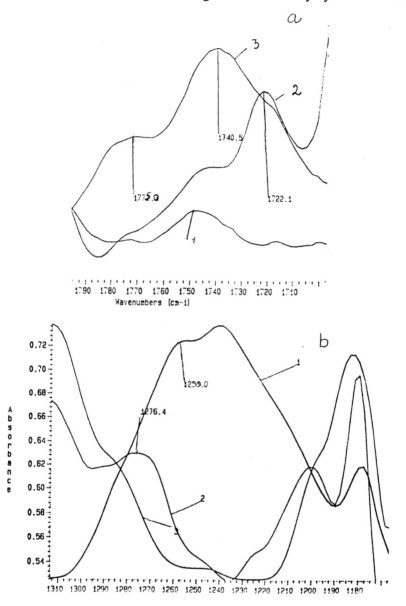

Figure 19. IR-spectrum PS2 at region 1700-1800 (a) and 1150-1310 cm-1 (b) a) original PS2 (1) and after thermal oxidative degradation 8.5 (2) and 215 min (3); b) original PS2 (3) and after thermal oxidative degradation 20(2) and 190 (1) min.

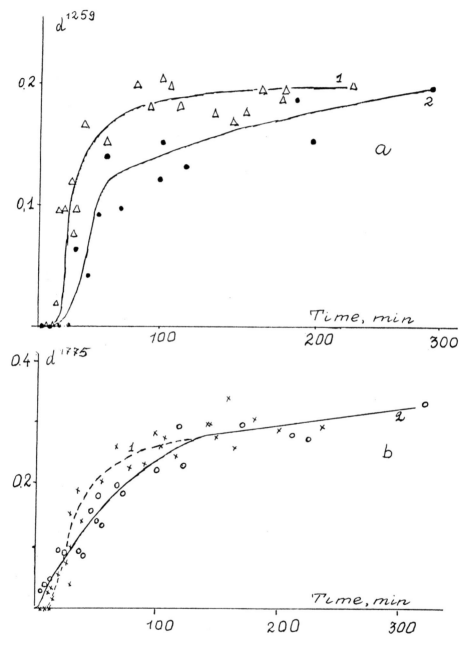

Figure 20. Kinetics of the accumulation bands at 1259 (a) and 1775 (b) of PS2 91) and PS2+IDBP (2).

The intrinsic mechanism of the inhibitory action of the additive is unclear. On the one hand, the additive is removed from polymer within the first 5 min of degradation (Figure 14); on the other the release kinetics of volatile products, which is noticeably slower in the presence of the additive (Figure 9), "feels" the latter within 300 min. Carbonyl compounds (Figure 18) are more rapidly formed and more rapidly depleted in the absence of the additive. Assuming that either processes occur with the participation of free radicals, the additive lower their concentration. Additive-induced deceleration is also observed for the kinetic of polymer residue weight changes (Figure 9) and for the bands 1375, 1450, 1495 cm^{-1} (Figures 15-17).

However, there are degradation products whose kinetics is not greatly affected by the additives. This seems to pertain to sufficiently stable oxidation products - ethers and acids (Figure 20).

In the kinetics of acid accumulation, there is autoacceleration with an induction period of 10 min, but in the presence of additive, the period of induction ceases. This phenomenon is similar to the action of homogenic catalysts, in the liquid-phase oxidation of hydrocarbons and associated with the higher rate constant of the intermediate-hydroperoxide.

By and large the experimental findings cannot provide unambiguous answers to what is the mechanism of the additive. However, they strongly suggest that the additive has an inhibitory action on the release kinetics of volatile products in the thermal oxidative degradation of polystyrene.

REFERENCES

1. G.G. Cameron & G.P. Kerr, *Eur. Polym. J.*, v. 4, p. 709 (1968).
2. I.C. McNeill, M. Zulfigar and T. Kousar, *Polymer Degr. and Stab.*, v. 28, p. 131 (1990).
3. J. Lucki and B. Ranby, *Polymer Degr. and Stab.*, v. 1, p. 273 (1979).
4. N. Grassie & N.A. Weir, *J. Appl. Polym. Sci.*, v. 9, p. 999 (1965).
5. M.C. Gupta & Y.D. Nath, *J. Appl. Pol. Sci.*, v. 25, p. 1017 (1980).
6. R.S. Lehrle, R.E. Peakman and J.C. Robb, *Eur. Polym. J.*, v. 18, p. 517 (1982).
7. J. Mullens, R. Garlier, G. Reggers, M. Ruysen, J. Yperman and L.C. Van Poucke, Bull. *Soc. Chim. Belg*, v. 101, N4, p. 267 (1992).
8. V.M. Goldberg, V.N. Yesenin and G.Ye. Zaikov, *Polymer Sci. USSR*, v. 28, N8, p. 1819 (1976).
9. M. Iring, M. Szesztay and A. Stirling, *Pure Appl. Chem.*, v. A29, N10, p. 865 (1992).
10. I.C. McNeill, W.T.K. Stevenson, *Polymer Degr. and Stab.*, v. 10, p. 247 (1985).

11. S.A. Danengauer, O.G. Utkina and Yu.N. Sazanov, *Journal of Therm. Anal.*, v. 33, p. 1213 (1988).
12. J. Dechant, R. Danz, W. Kimmer, R. Schmolke, Ultrarotspektroskopische untersuchungen an olymeren, Akademie-Verlag-Berlin (1972).
13. E.T. Denisov, Yu.B. Shilov, *Vysokomolek. soed.*, v. 25A, N6, p. 1196 (1983).
14. V.N. Yesenin, I.A. Krasotkina, G.Ye. Zaikov, *Vysokomolek. soed.*, 25B, N1, p. 48 (1983).

KINETIC MODEL OF POLYMER DESTRUCTION OCCURRING DURING EXTRUSION PROCESS

E.G. Eldarov and F.V. Mamedov
The State Azerbaidjan Academy of Oil, Baku city, Azerbaidjan

V.M. Goldberg and G.E. Zaikov
Institute of Chemical Physics named after Semenov,
Russian Acad. of Sci., Moscow, Russia

INTRODUCTION

In this paper a model is given describing the interrelation of mechanical destruction and thermal oxidation phenomena during PE extrusion. This model permits to determine changes in molecular weight, the amounts of reacted oxygen and the quantities of accumulated peroxide together with the amount of the inhibitor used during extrusion. These parameters produce a significant effect on the physical and mechanical properties of the polymer and make it possible to assess its lifetime in finished products under operating conditions.

Polymer processing is usually accompanied by numerous chemical reactions which end up in transformations leading to considerable changes in physical and mechanical properties of polymer materials causing reduced lifetimes of finished products.

The chemical aspect of this problem is hard for numerical representation. Relevant chemical reactions take place at relatively high temperatures due to differing initiation sources. During extrusion processing certain thermal, mechanical and autocatalytic reactions occur involving free radicals, ions, ionic pairs and low molecular weight combinations. The current status of chemistry of polymer destruction does not describe so complicated kinetics of decomposition process. Practical needs, however, require mathematical models which can address changes occurring in polymer materials [1]. Existing models do not take into account chemical transformations in the central link of the chain. These transformations are influenced by the specific reaction conditions and in turn determine changes in polymer's structure. It is clear that in such cases the possibilities of extrapolating beyond the investigated conditions are all but limited.

In this paper a tentative to describe transformations in polymers during their processing is made by way of examining the kinetics of relevant reactions, i.e. a step aimed at improving the quality of the destruction model has been made. During the processing at sheared rates of 10-10^2 sec $^{-1}$, temperatures of 180-250°C and in presence of oxygen certain mechanical and thermal oxidation transformations occur. As an example the behavior of regular PE during extrusion has been studied.

Destruction is manifested through the formation of alkine R° radicals which interact with oxygen to form peroxide RO_2. The formation of RO° radical will eventually lead to breaking of macromolecular links and decreasing of molecular weight. The molecular weight is the most important parameter which should obviously be used to optimize the processing conditions.

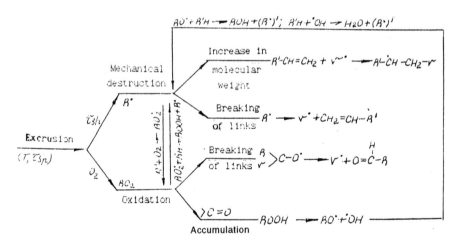

Figure 1. Changes in macromolecules occurring during polymer processing.

The technique of determining the molecular weight during polymer's processing has been described and verified in paper [2], Figure 1. The results of this research work indicate that the kinetics of chemical transformations taking place during polymers' processing can be described by a regular thermal oxidation reaction in which the primary mechanically-initiated break in the polymer's chain is determined by the W_0 parameter. Above-mentioned paper [2] analyzes certain postulates stating that W_0 does not have a direct bearing on the decrease of molecular weight, the latter can actually increase or decrease while W_0 determines only the concentration of free radicals involved in both processes. It is evident that W_0 is the function of temperature, rheological parameters and molecular weight but the studies of these interrelations are not the subject matter of this paper.

According to the scheme of macromolecule transformation given in Figure 1 there are two reasons for the decrease of molecular weight: isomerization process leading to breaking of alkyne radical $R°$ (Reaction $2_d^{RO°}$) and decomposition of alcoxyl radical $RO°$ (Reaction $2_d^{RO°}$). The increase of molecular weight is achieved through the joining of $R°$ radical to the double bond (Reaction 2_i^{RO}). These three processes take place simultaneously and rival each other. There are two more reactions causing mechanically-initiated bond breaking (Reaction 0 and Reaction 4). They produce certain impact on the size of macromolecules but may be neglected because the reaction rates are of the same magnitude while their directions are opposite.

Quantitative representation of the autooxidation of the polymer's melt at various oxygen concentrations and in presence of an inhibitor seems to be an important aspect of the entire problem. The basic data required for such calculations is contained in literature [3-5].

At first phases during the development of the model the number of elementary steps can be limited. The diffusion problems that might happen during polymer's mixing and processing can be neglected.

The kinetics of chemical transformations is the following:

0. $R - CH_2 - CH_2 - B' \xrightarrow[T,\overline{u}gb]{W_o} R'CH_2 + \cdot CH_2R''$ Breaking of carbon chain -

                                                    initiation

1. $R + O_2 \xrightarrow{K_1} RO_2$ Developing of kinetic chain

2. $RO_2 + RH \xrightarrow{K_2} ROOH + R$ Branching of kinetic chain

3. $ROOH \xrightarrow{K_3} RO + OH$

2. $R + R'H \xrightarrow{K_2^{R'}} RH + (R')$

2. $RO \cdot + RH \xrightarrow{K_2^{RO'}} ROH + R \cdot$ Developing of kinetic chain

2. $HO^{\cdot} + RH \xrightarrow{K_2^{OH}} H_2O + R^{\cdot}$

$$2. \quad R'''{-}CH - CH_2 - R'''' \xrightarrow{K_\alpha^{RO}} R'''{-}\overset{H}{\underset{O}{C}} + {\overset{\cdot}{C}}H_2 - R'''' \qquad \text{Breaking of carbon}$$

chain

2. $R'''CH_2 - \overset{\cdot}{C}H - CH_2 - R'''' \xrightarrow{K_\alpha^{B}} B'''CH = CH_2 + {\overset{\cdot}{C}}H - B''''$

2. $R'''CH = CH_2 + R^{\cdot} \xrightarrow{K_i^{B}} R'''{-}\underset{R}{CH} - {\overset{\cdot}{C}}H_2$ Developing of carbon chain

4. $R^{\cdot} + R^{\cdot} \xrightarrow{K_4} R - R$

5. $RO_2^{\cdot} + R^{\cdot} \xrightarrow{K_5} ROOR$ Breaking of kinetic chain

6. $RO_2^{\cdot} + RO_2^{\cdot} \xrightarrow{K_6}$ Alcohol+ketone+oxygen

7. $RO_2^{\cdot} + InH \xrightarrow{K_7} ROOH + In^{\cdot}$ Transfer of kinetic chain

8. $RO_2^{\cdot} + I^{\cdot}n \xrightarrow{K_8}$ Molecular product   Breaking of kinetic chain

10'. $I^{\cdot}n + RH \xrightarrow{K_{10}}$ Molecular product $+R°$ Transfer of kinetic chain

10." $InO_2^{\cdot} + RH \xrightarrow{K_{10}''} InOOH + R^{\cdot}$

11. $In^{\cdot} \xrightarrow{K_{11}} InO_2^{\cdot}$

The above combination of elementary steps should be regarded as rough approximation to the real chemical process. For example, Reaction 3 is one of the many reactions relating to the mechanism of peroxides decomposition. It is known that peroxides can decompose through a bimolecular process which involves another peroxide group, "C-H" bond or free radicals. The destruction can continue down to molecular products. Besides, the peroxide decomposition constant can depend upon the nature of adjacent functional group, i.e. hydroxyl, carbonyl or carboxyl. The above approach is especially useful for describing inhibitor's behavior which may include several phases. The decision of limiting the number of phases looks to a large extent arbitrary. In these circumstances the authors decided to utilize their own experimental data showing that at high oxidation temperatures the number of elementary phases sufficient for description of this complex process shall not be large. It is explained by the fact that under such conditions the reactions having highest activation energies prevail. The above indicated kinetic scheme permits to compose a system of differential equations representing polymer's oxidation, the amounts of inhibitor used and the change of molecular weight. Omitting the intermediate calculations we shall indicate below only the final expressions describing the kinetics of changes in concentrations of major oxidation products. For convenience the time differentiation was replaced by the differentiation against l - length of extruder's screw.

$$-\frac{d[O_2]}{d\ell} = V^{-1}\left[(K_2 RH + K_7[InH]) \cdot \sqrt{\frac{K_3[ROOH] + W_o}{K_6\left(\frac{K_7 \cdot K_8[InH]}{K_6 \cdot K_{10} RH} + 1\right)}} \middle/ \left(1 + \frac{K_2 RH \sqrt{K_4}}{\sqrt{K_6} \cdot K_1[O_2]}\right)\right.$$

$$\left. + K_3[ROOH]\right]\ldots$$

$$(1)$$

$$\frac{d[ROOH]}{d\ell} = V^{-1}\left[(K_2 RH + K_7[InH]) \cdot \sqrt{\frac{K_3[ROOH] + W_o}{K_6\left(\frac{K_7 \cdot K_8[InH]}{K_6 \cdot K_{10} RH} + 1\right)}} \middle/ \left(1 + \frac{K_2 RH \sqrt{K_4}}{\sqrt{K_6} \cdot K_1[O_2]}\right)\right.$$

$$\left. - K_3[ROOH]\right]\ldots$$

$$(2)$$

$$-\frac{d[InH]}{d\ell} = V^{-1}\left(\frac{K_7[InH] \cdot \sqrt{K_3[ROOH] + W_o}}{\sqrt{K_6\left(\frac{K_7 \cdot K_8[InH]}{K_6 \cdot K_{10} RH} + 1\right)}} \middle/ \left(1 + \frac{K_2 RH \sqrt{K_4}}{\sqrt{K_6} \cdot K_1[O_2]}\right)\right)\ldots$$

$$(3)$$

Whereas,
V - linear speed of melt's flow in the extruder's barrel (or its projection for the extruder's length);

$$\left(\frac{K_7 \cdot K_8'[J_n H]}{K_6 \cdot K_{10} RH} + 1\right)$$ - parameter which takes into account the input of the inhibitor into bond breaking.

$\left(\dfrac{K_2 RH \sqrt{K_4}}{\sqrt{K_6 K_1 [O_2]}} + 1 \right)$ - parameter which takes into account $R^{\cdot}$ to $RO_2^{\cdot}$ rela-

tions as well as the input of $R^{\cdot}$ radical into the breaking of kinetic chains.

$K_{10} = K_{10}{}' + K_{10}{}''[O_2]$ - parameter which takes into account the reaction of chain development insured by both "alkyne" $J_n{}'$ and peroxide $JnO^{\cdot}2$ inhibitor radicals.

The solution of these equations makes it possible to obtain concentration profiles for oxygen, peroxide and the inhibitor (along the extruder's length).

It should be noted that W_0 parameter, being the function of temperature and shear rate, can also vary lengthwise but for the first approximation its mean value can be considered as constant.

The calculation of alkyne $R^{\cdot}$ radical and alcoxyl $RO^{\cdot}$ radical concentrations as well s the increase of polymer's chain does not present any difficulty:

$$[RO_2^{\cdot}] \approx \dfrac{\sqrt{K_3 [ROOH] + W_o}}{\sqrt{K_6 \left(\dfrac{K_7 \cdot K_8 [J_n H]}{K_6 \cdot K_{10} \cdot RH} + 1 \right) \left(1 + \dfrac{K_2 RH \sqrt{K_4}}{\sqrt{K_6 \cdot K_1 [O_2]}} \right)}} \dots \tag{4}$$

$$[R^{\cdot}] = \dfrac{K_2 [RH]}{K_1 [O_2]} [RO_2^{\cdot}] \dots \tag{5}$$

$$[RO^{\cdot}] = \dfrac{K_3 [ROOH]}{K_\alpha + K_\alpha^{RO^{\cdot}} [RH]} \dots \tag{6}$$

In this case the change in molecular weight resulting from specific conditions of the extrusion process can be represented as the competition of decomposition reactions $(2_d^R$ and $2_d^{RO^{\cdot}})$ and the increase of the size of molecular chain (2_i^R).

$$\dfrac{dn}{dt} = \left(K_d^R - K_i^R \right)[R^{\cdot}] + K_d^{RO} [RO^{\cdot}] \dots \tag{7}$$

n_t value, being the change of macromolecules in a unit of weight of polymer material (one kilogram) is represented through the molecular weight as follows:

$$n_t = 1 \cdot 10^3 \left[\frac{1}{(M_n)_t} - \frac{1}{(M_n)_o} \right] \dots \tag{8}$$

then, during the extrusion

$$\frac{dn_\ell}{d\ell} = -1 \cdot 10^3 \frac{d(M_n)_\ell V}{(M_n)^2 d\ell} \dots \tag{9}$$

$$-\frac{d(M_n)_\ell}{d\ell} = \frac{(M_n)^2}{1 \cdot 10^3 V} \left(K_d^R - K_i^R \right) [R^\cdot] + K_d^{RO} [RO^\cdot] \dots \tag{10}$$

This model can be helpful for the calculations of possible changes of molecular weight and the amount of absorbed oxygen. These two parameters produce a significant influence on physical and mechanical properties of the polymer material. No less important is the information about the quantities of accumulated hydroperoxide and amounts of inhibitor used during processing. This data will permit to assess the lifetime of the polymer under operating conditions. Thus, the proposed model is capable of not only describing the changes of properties but also predicts the lifetime of the processed material.

When developing such a complex model it is highly important to properly determine the basic values of kinetic parameters. Some of these parameters were many times determined at low temperatures, namely $K_2 \cdot K_6^{-1/2}, K_7 \cdot K_6^{-1/2}$.

The scatter of activation energies as well as preexponents of these parameters described elsewhere is relatively small and does not permit their arbitrary alteration within a wide range.

Other values utilized in calculations, for example reaction rate constant relating to bonding of O_2 to alkyne radical K_1, the rates of quadruple recombination of alkyne (K_4) and peroxide (K_6) radicals as well as cross reactions of inhibitor $J_n^\cdot$ and peroxide $RO_2^\cdot$ - (K_8) used to be measured rather seldom. Their magnitudes are high but they vary within a narrow range. Papers [3,5] contain experimental values of the constants of kinetic chain transfer to inhibitor's radicals both without the involvement of (K_{10}') and with the involvement of (K_{10}'') of oxygen.

Little data is available about the constants of rates of alkyne radical: for decomposition (K_d^{RO}) and for chain development (K_2^{RO}). In

132   E.G. Eldarov et al.

our case, however, the main contribution to the calculations is done not by the absolute value of these constants but by the relation of the constant of decomposition rate to the resulting constant of the absorption of RO˙ radical. In its value this relation is close to the opposite number of oxygen molecules needed to break one bond. For high temperatures it is within the range of 3-15. In contrast, extensive data is available about the decomposition constant of PE hydroxide [K₃], but the decomposition mechanism is not quite clear that is why a simple monomolecular decomposition at high processing temperature has been used in this research.

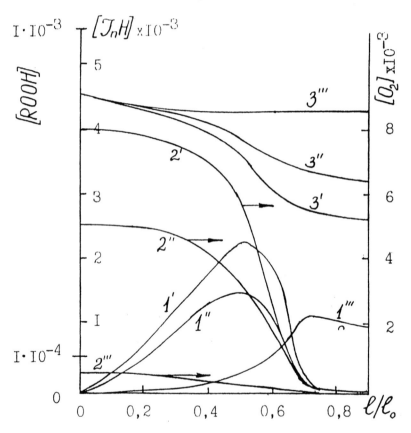

Figure 2. Profiles of peroxide [ROOH] (1', 1", 1'''), oxygen [O₂] (2', 2", 2''') and the inhibitor [JₙH](3', 3", 3''') as per extruder zones. Calculating conditions are the following: (') - [O₂]-8·10⁻³; (") - [O₂]-5·10⁻³; (''') - [O₂]-0,5·10⁻³; for all cases W₀-1·10⁻⁴; (K₂)₀-1·10⁹; (K₃)₀-5,4·10¹⁶.

This decomposition process is characterized by a high activation energy which may constitute 35 kcal/mol and by a preexponent exceeding $1 \cdot 10^{13}$ sec^{-1}.

Practically, there is no information about W_0 (the rate of thermal oxidation and mechano-chemical initiation of radicals). The upper limit of this value, $W_0 = 1 \cdot 10^{-3}$ mol/kg·sec has been chosen due to the following reasons: at certain process conditions the inhibitor can be fully absorbed at its initial concentration of $[J_nH]_0 \approx 1 \cdot 10^{-2}$ mol·OH/kg. Assuming that the extrusion time is approximately $1 \cdot 10^2$ sec and the excess of the rate one decimal order of magnitude, then the above-mentioned W_0 value can be achieved.

Figure 2 shows oxygen concentration (1', 1") peroxide concentrations (2', 2") and inhibitor concentrations (3', 3"). Calculation conditions are the following: $[O_2] = 8 \cdot 10^{-3}$ and $5 \cdot 10^{-3}$, $[K_2]_0 = 1 \cdot 10^9$ kg/mol·sec, $[K_3]_0$- $5,4 \cdot 10^{16}$ sec^{-1}, W_0-$1 \cdot 10^{-4}$ mol/kg·sec.

Temperatures as per extruder's zones vary within the range of 150-220°C. It was discovered that the consumption of oxygen at the entry side of the extruder is practically unnoticeable (this happens at 1/5th of extruders length). Beyond this section oxygen consumption gradually increases and at the point of 0,7 of the extruder's length all oxygen is consumed. Concentration of hydroperoxide attains maximum and then drops to zero. After similar period of time the inhibitor stops being utilized.

Profiles of key oxidation products are not virtually dependent on oxygen concentration. However, this concentration influences absolute values of maximal hydroperoxide concentration as well as the amount of inhibitor used during the extrusion process.

An important feature of the calculations is the availability of an extruder's zone in which there is no oxygen and hence there is no oxidation. Such processing conditions might be needed to obtain a product with an increased stability to thermal oxidation (because in such a case the content of peroxide is zero). During processing of secondary polymers (or used polymers) which accumulated peroxide during their previous use it is needed to decrease peroxides' concentration. In spite of theoretical and practical feasibility of creating an "embellishing" zone within the extruder's barrel there have been no actual investigations to this effect. No experimental data is available on this subject.

Figure 3 shows peroxide profiles at differing W_0 values and various shear rates. Thermal oxidation is somewhat accelerated comparing to the conditions depicted by Figure 2 particularly, decomposition constant of the peroxide is higher by an order of magnitude. This resulted in the shift of peroxides primarily at the entry side of the extruder. It

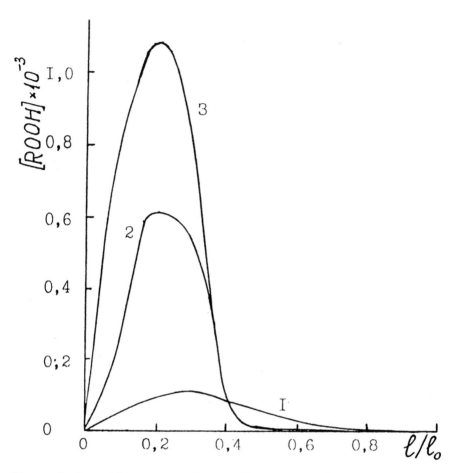

Figure 3. Peroxide concentration profiles per extruder zones at various rates of primary initiation: 1. $W_0\text{-}1\cdot10^{-5}$; 2. $W_0\text{-}1\cdot10^{-4}$; 3. $W_0\text{-}1\cdot10^{-3}$. Calculating conditions are the following: $[O_2]$ - $1\cdot10^{-2}$; $(K_3)_0\text{-}5{,}4\cdot10^{-17}$.

was assumed (and the assumption happened to be true) that oxygen concentration profile will determine inhibitor's and peroxides' concentrations. Therefore, in future the authors will concentrate on the problem of oxygen absorption along the extruder's length.

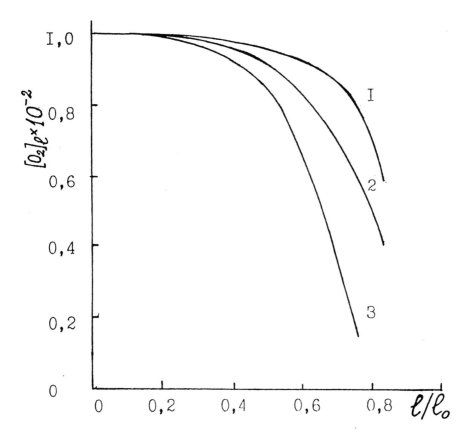

Figure 4. Oxygen absorption kinetics versus $(K_2)_0$: 1. $(K_2)_0=0,27 \cdot 10^8$; 2. $(K_2)_0-0,8 \cdot 10^8$; 3. $(K_2)_0-2,7 \cdot 10^8$.

Figure 4 shows oxygen profiles at different constants of the rate of kinetic oxidation chain development. The increase of these constants means a transfer from less oxidizable polymers to more reactive, for example, from PEVP to PP.

A specific feature of such calculation as well as other relevant calculations is that they are not conducted till the very end but are finished at the phase when a rapid increase of oxygen adsorption occurs. If calculations are continued beyond this point the computer time will greatly increase (the data yield will not however increase significantly). It is obvious that oxidation process will not go any further.

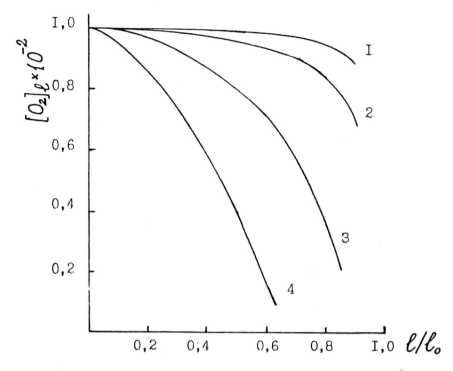

Figure 5. Oxidation rate versus initiation rate $[W_o]$. 1. $1 \cdot 10^{-6}$; 2. $1 \cdot 10^{-5}$; 3. $3.1 \cdot 10^{-4}$; 4. $1 \cdot 10^{-3}$.

As it is seen from Figure 4 the increase of $(K_2)_o$ - permanent oxidation rate brings about the shift of the starting point of the acceleration towards the loading zone of the extruder. This in turn increases the size of the embellishing zone. Similar results were obtained with the increase of W_o value, i.e. the capacity of the reactor which resulted in a higher shear rate, $\dot{\gamma}$, and therefore higher rate of initial mechanically-initiated breaking of macromolecular.

Figure 5 shows oxygen profiles at W_o values varying from $1 \cdot 10^{-6}$ up to $1 \cdot 10^{-3}$ mol/kg·f. At $W_o < 4 \cdot 10^{-5}$ mol/kg·f the influence of W_o becomes small. When W_o becomes $\geq 4 \cdot 10^{-5}$ mole/kg·f it starts to produce a more significant impact on the whole process than the autooxidation process itself. It should be noted that the gap between Curve 2 and Curve 4 (minimal and maximal influence of W_o, respectively) is rather narrow. The actual gap for $\dot{\gamma}$ values can be still narrower because it is highly probable that W_o will increase exponentially with increasing. However, the reduction of polymer's residence time shall be the reason of an opposite phenomenon, namely the increase of the range of "functional values" of $\dot{\gamma}$.

The behavior of the inhibitor can be regarded as a critical factor decisive for the rate of oxygen absorption. It is common knowledge that the higher is the inhibitor's content the better is the stability of polymer during extrusion. This is the reason why PE insulations contain up to 5% of antioxidant when they are used for critical applications.

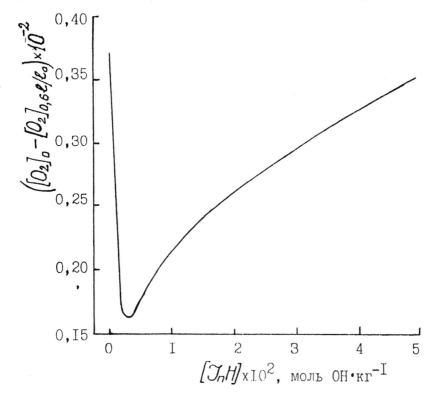

Figure 6. Mean oxygen absorption rate versus initial inhibitor's concentration.

Studies of the mechanism of inhibiting autooxidation conducted recently have shown that at high temperatures the inhibitor acts as an initiator of the oxidation process. The model described in this paper takes this phenomenon into account by describing the transfer of kinetic chain with and without the involvement of oxygen. Figure 6 shows the amounts of absorbed oxygen versus inhibitor's concentrations. As it is illustrated at low inhibitor's concentrations the amount of absorbed oxygen is low but it tends to gradually increase once inhibitors concentrations get higher. The action of the inhibitor becomes self eliminating at $[J_nH] \approx 4,510^{-2}$ mol OH/kg, i.e. at 1% for "space limited" phenols. During this phase the oxidation rates of inhibitor-doped and inhibitor-free polymers become the same. This simplified approach of describing inhibitor's action indicates that there is a certain optimal inhibitor's concentration which ensures maximal retardation of polymer's oxidation. If represented by IONOL or NONOX WSP this concentration can constitute less than 0,1% by weight.

Investigations aimed at determining optimal stabilization conditions for thermoplastic materials shall have to be supplemented by the results of recent studies conducted in the field of kinetics of inhibited oxidation.

It is also important to take into account such phenomena (that have been recently recognized) as the ramp of inhibited oxidation at increasing inhibitor concentrations. High effective activation energy of oxidation and low solubility of oxygen in the crystallic fraction of semicrystallic polymers are the indicators of high oxidation rates taking place during polymer processing. These rates are higher than those occurring normally under operating conditions. Therefore, it is more important to stabilize PE during its processing than to try to reduce PE oxidation when it is under operating conditions. When polymer is used in finished products other types of destruction would cause a reduction in PE's lifetime, namely photooxidation and mechanical destruction.

Polymer macromolecules' behavior during its processing acquires a particular importance. Equation 10 shows that destruction going at W_O rate can be caused by a mechano-chemical initiation process which will eventually result in RO° radical decomposition, isomerization and breaking of R° at a rate constant being K_d^{Ro}. The growth of macromolecules is explained by the fact that $R^{\bullet}$ radicals are joined to the end double bond.

The growth of macromolecules does not take into account the recombination of $R^{\bullet}$ radicals because at oxygen concentrations of $[O_2] \geq (5 \cdot 10^{-5} - 1 \cdot 10^{-4})$ mol/kg the kinetic importance of this reaction is small. This group of reactions is noted by the availability of a considerable difference in their activation energies. W_O value, being the kinetic characteristic of the mechano-chemical process, does not have to possess a tangible temperature coefficient and should be defined arbitrary for every particular regime. Efficient activation energies of joining $R^{\bullet}$ radical to the double bond as well as isomerization energy of R° radical are 4-6 and 15-20 kcal/mol, respectively, i.e. essentially different and substantially larger than W_O value. RO° radicals decomposition (this radical is derived from peroxide) is characterized by a still higher temperature coefficient of destruction, which value is 30-35 kcal/mol.

Therefore, the changes happening in macromolecules are expected to be in a very complex relation with the temperature of polymer's processing. The behavior of this parameter can not be predicted for polymer's processing because temperature keeps varying along the length of extruder's barrel. To get a better understanding of relevant interdependences all the calculations connected with the macromolecules are indicated at a constant temperature for every separate processing regime.

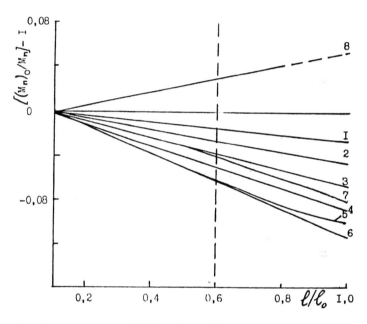

Figure 7. Scissions (crosslinkings) profile along extruders at the temperatures: 1-140°C, 2-150°C, 3-160°C, 4-170°C, 5-180°C, 6-190°C, 200°C, 7-210°C, 8-220°C (calculation conditions: $W_o=2 \times 10^{-6}$ $(K_2)_o=2,7 \times 10^7$, $(K_3)_o=5,4 \times 10^{13}$, $(K_d)_o=1 \times 10^{10}$, $(K_i)=1 \times 10^6$.

Figure 7 shows profiles of breakages of cross-links along the extruder at differing temperatures at minimal and practically negligible W_o - $2 \cdot 10^{-6}$ mol/kg·f. The curves given in Figure 7 are of autoaccelerative nature. This proves that the oxidation reaction seems to be dominating over other chemical reactions resulting in changes in macromolecules. At lower temperatures the macromolecules tend to grow, hence their concentration decreases. Up to 190°C temperatures this effect seems to increase. At temperatures of 210-220°C a transition from macromolecules' growth to their shrinking occurs.

The collected data proves that the thermal oxidation reactions will lead to two types of changes in macromolecules: low temperature growth and high temperature destruction.

It should be noted that at acceptable constant value of peroxide decomposition, K_3, and constant oxidation rate dependent on the constant of kinetic chain growth, K_2, the entire input into the destruction process will be negligible. In fact, only two competing reactions of $R^\bullet$ radical should be studied: the reaction of joining $R^\bullet$ radical to the double bond and secondly the isomerization reaction which ends up in a break. The maximal stability of the material can be obtained at temperatures close to 220°C. The correctness of this assumption was proved by the experimental results. Figure 8 shows a mean number of cross-links' breaks, calculated per one molecule, S - $[(M_n)_0 / M_n]$ - 1 at differing processing temperatures and W_0 values. When temperature rises from 150 to 180°C the macromolecules tend to grow at a receding rate. At 190°C temperature the growth is of the same magnitude as at 180°C temperature. Further elevation of temperature would slow down the growth which eventually becomes zero at 210-220°C. At still higher temperatures destruction of the polymer material starts. Such phenomenon was observed when testing low density PE in the atmosphere of inert gas in rotary-type viscosimeter and also when imitating operating conditions by utilizing an extrusion meter as well as a device called conteriograph [48].

Relative number of cross-link breaks versus temperature at midpoint of extruders barrel is shown in Figure 9. Low temperature destruction seems to start at W_0 - $2 \cdot 10^{-5}$ mol/kg·f (curve 2). It is noteworthy that the right side of the curve is steeper than the left side.

Macromolecules grow at temperatures varying from 150 to 215°C. When W_0 ramps up to $2 \cdot 10^{-4}$ mol/kg·f (curve 3) the changes occurring in macromolecules become destructive. It is understandable because radicals' concentrations increase, i.e. the rate of the process and the balance of reactions tends to move to destruction. The domain of macromolecules' growth gets significantly smaller both with regard to the temperature range (170-195°C) as well as the number of cross links per molecule (0,05 instead of 0,15).

After conducting experiments similar results were obtained. At low values of W_0 no low temperature destruction was noticed. At high W_0 (when testing material in disc extruder) no cross linking was observed whatsoever [92]. Temperature interface between destruction and growth was similar to the data obtained experimentally.

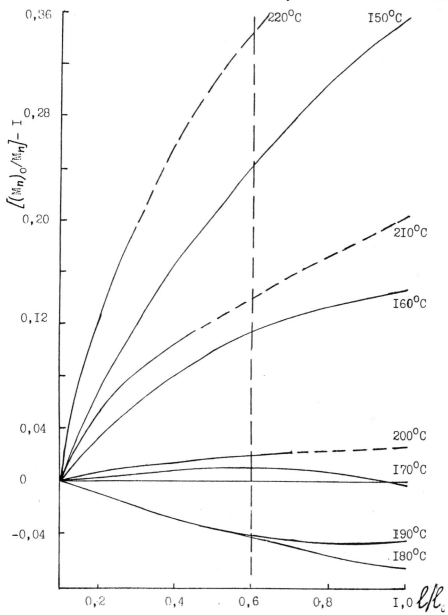

Figure 8. Profile of macromolecules' breaks at differing temperatures along the extruders length. Calculating conditions: W_0-$2 \cdot 10^{-4}$; $(K_2)_0$-$2,7 \cdot 10^7$; $(K_3)_0$-$5,4 \cdot 10^{13}$; $(K_d)_0$-$1 \cdot 10^{10}$; $(K_i)_0$-$1 \cdot 10^6$.

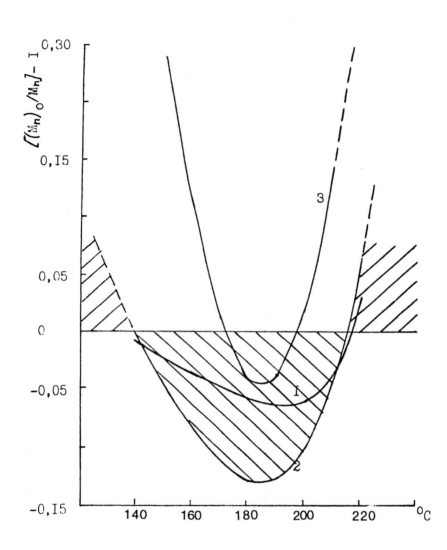

Figure 9. Relative changes occurring in macromolecules of the midpoint of extruder's barrel. 1. W_o-$2 \cdot 10^{-6}$; 2. W_o-$2 \cdot 10^{-5}$; 3. W_o-$2 \cdot 10^{-4}$.

A particular attention should be paid to the role that the rate of oxidation plays in the entire process of macromolecules' change. At high W_o values the oxidation process goes independently from mechanical destruction. Curve 2 shows that similar changes in macromolecules can happen at temperatures differing by 40-50°C (for example, when the number of breaks S - -0,05). Oxidation rates in such cases are quite different. It can be seen that mechanical destruction prevails over thermal oxidation. As it was stated before, the general notion of what prevails - oxidation or mechanical destruction - does not seem to have a "chemical meaning." However, this question should not be regarded as purely abstract or academic because it has a practical importance. Depending upon the concentration of oxygen and interrelation of chained and non-chained processes they should be retarded either by means of antioxidants or by using acceptors of $R^{\bullet}$ radicals. It is obvious that in situations when mechanical destruction prevails (high W_o values) the use of antioxidants will be less efficient. In fact, during extrusion one should use compounds comprising antioxidants, peroxide decomposers and acceptors of alkine radicals.

CONCLUSIONS

1. As it was stated in this paper there shall be an extruder's zone in which there is no oxygen in the polymer material, i.e. no oxidation, no inhibitor and no peroxide. Such a zone can be called an "embellishing" zone.

2. To obtain the required stabilizing effect the inhibitor's concentration shall be optimal. During normal extrusion it should be $[J_nH]_{opt}$ - $3 \cdot 10^{-3}$ mol OH/kg: that is substantially lower than the concentrations currently used by polymer industry.

3. There are three temperature-dependent processes that result in shrinking or growth of macromolecules:

low temperature destruction occurring due to mechanically initiated breaks;

medium temperature macromolecule growth happening because alkine $R^{\bullet}$ radicals are joined to macromolecules double bonds;

high temperature destruction occurring due to isomerization which leads to breaking of macromolecules' chains. The influence of the above phenomena is dependent on the properties of polymer being processed and on processing conditions.

REFERENCES

1. MG. Hametova, V.S. Kim, Plastko-87, *Summaries Book*, Dum Techniky, CSVTS, 1987, z. 13-14.
2. V.M. Gol'dberg, G.E. Zaikov, *Polymer Degradation and Stability*, 1987, 19, p. 221.
3. V.M. Gol'dberg, L.A. Vidovskaya, G.E. Zaikov, *Polymer Degradation and Stability*, 1988, 20, p. 93.
4. V.M. Gol'dberg, G.E. Zaikov, *Polymer Yearbook*, 1988, 5.
5. P.S. Belov, V.M. Gol'dberg, G.A. Vidovskaya, *VMS*, 27A, 1985, p. 2048.
6. E.T. Denisov, *Oxidation and Destruction of Carbochain Polyimides*, Chemistry, 1990 , p. 218.
7. Yu.A. Shlyapnikov, S.G. Kiryushkin, A.P. Marin, Antioxidation and stabilization of polymers, Chemistry, 1986, p. 252.
8. E.F. Brin, O.N. Karpukhin, V.M. Gol'dberg, Opposite task of chemical kinetics at determining of the mechanism of inhibited high-temperature PE oxidation, *Chemical Physics*, 1986, 5, N7, pp 938-947.
9. T.A. Bogaevskaya, N.K. Tyuleneva, Yu.A. Shlyapnikov. *Oxidation of PE stabilized by a strong phenol antioxidant. VMS*, 1981, 23A, N1, pp. 181-186.
10. T.V. Monachova, T.A. Bogaevskaya, Yu.A. Shlyapnikov, *Inhibited oxidation of rigid crystallic polypropylene, VMS*, 1974, 16B, N11, pp. 840-841.
11. L.A. Tatarenko, Yu.A. Shlyapnikov. *Processes happening during the induction period of inhibited polyethylene oxidation, VMS*, 1990, 32A, pp. 2318-2323.
12. V.M. Gol'dberg, V.M. Yarlykov, N.G. Paverman, M.S. Akutin, E.I. Berezina, G.V. Vinogradov, *VMS*, 1978, A20, p. 2437.
13. S.V. Portnenko, V.B. Dvornichenko, P.I. Bashtannic, E.A. Sporjgin, *Voprosy Chimii i Technololgii*, Charcov, 1989, N91, p. 611.

PROBLEMS OF AGING AND SEARCHING WAYS FOR POLYVINYLCHLORIDE STABILIZATION

K.Z. Gumargalieva, V.B. Ivanov, G.E. Zaikov, Ju.V. Moiseev, and T.V. Pokholok

Institute of Chemical Physics, 4 Kosygin str., Moscow 117334, Russia

Abstract— There was performed complex investigation of chlorinated polyvinyl plasticate samples, aged in model and climate conditions, and the ones been exploited for a long time (15-30 years at 253-301 K). The mechanism of their aging was found.

It was shown that the loose of exploitational and other functional properties of PVC-plasticate occurs, first of all, as a result of desorption of plastifier and other additives. The following methods were applied for this purpose: thermogravimetry, chromatography, UV- and IR-spectroscopy, gel-penetrating chromatography and mercuric porometry, measurement of stability characteristics of the materials (for example, σ_s and ε_b at break).

Predominant process for PVC-plasticate at its dark low temperature aging is plastifier desorption, and at light one- degradation of polymer and plastifier, proceeding, generally, on the irradiating sample side and depending on spectrum composition of the light falling. There occurs washing off of thermostabilizers from the material at the contact with drop water. With regard to experimentally determined empyric kinetic equations of mass losses or expensing additives during exploitation or exposure, one can determine the conditions of material aging. Basing on the analysis of exploitation conditions, one can deduce particular prognosing equation for the description of PVC-plasticate degradation proceeding.

INTRODUCTION

The main factors causing PVC-plasticate aging at storage and exploitation of materials or articles, are continuously influencing temperature, oxygen, humidity, mechanical stresses, aggressive media and ionizing radiations, leading to the change of initial exploitation properties, resulting simultaneous proceeding chemical and physical processes, changing chemical composition and structure of the material. In general, the known investigations relate to the study of one of the processes, high temperature degradation, for example [1-3]. The complex of processes, proceeding in PVC-plasticate, is reduced to dehydrochlonation reactions, leading to HCl detachment with the formation of polyene regularities, their amount and length defining material color; to thermooxidation degradation with the formation of carbonile-containing groups; diffusional desorption of plasticizer and water sorption. In real PVC-plasticate aging or any other polymer material not all processes will proceed. Moreover, the proceeding of just a part of them will lead to exploitation properties change.

That is why, at the study of any polymer material aging it is necessary to determine aging model, representing totality of kinetic data as establishment of aging processes, proceeding in polymer, responsible for the change of exploitation properties and obtaining formally-kinetic equation, connecting indexes of exploitation nature with time and aging factors [1-4]. In this case selected aging model does not allow to make a conclusion about full aging mechanism, but is enough for the description of kinetics of polymer material aging in particular exploitation conditions.

It is necessary to use characteristic aging indexes, reflecting fundamental material properties (chemical composition, molecular-weight distribution, physical structure), independent on surrounding conditions, in order to differ aging from the change of polymer properties, caused by the change of external medium conditions (temperature, humidity). From this point of view probably, the main process of PVC-plasticate again at room temperature is diffusional plasticizer desorption. At temperatures over 60°C it is probable the proceeding of thermooxidation degradation, which may be accompanied by dehydrochlorination reaction. It is advisable to control the proceeding of the first process by residual amount of plasticizer in PVC. The presence of thermoxidation degradation and dehydrochlorination may be controlled by the change of molecular-weight polymer distribution and by accumulation of double bonds in it.

THE OBJECT AND INVESTIGATION METHODS
FOR AGING KINETICS OF PVC-PLASTICATE

There were investigated wire isolation, produced from PVC-plasticate of M64 mark (USSR), exploited during a long time period (up to 28 years) in aircraft engines, and received from industry after 1 or 2 year storage, and PVC films with various content of dioctylphtalate and dialkylphtalate.

Plastifier output was calculated according to empyric equation [4]:

$$m_t = kt^n, \qquad (1)$$

where m_t - plastifier amount, detached during time t; n and k - parameters ($0.5 \leq n \leq 1$).

To determine residual amount of plastifier in CPV thermogravimentric, chromatographic and UV-spectrophotometric methods were used.

Thermoanalyzers 1090 of DuPont Company and TA 3000 of Mettler Company were applied for thermal analysis performance. Measurements were performed in both isothermal and dynamic regime at variation of heating rate in 0.1-100 K/min range.

UV-spectra of PVC specimen solution in tetrahydrofurane were investigated on UV-spectrophotometer PS 800 of Pye Unicam Company in 185-700 nm range.

Chromatographic analysis was performed on gas-fluid chromatograph of "color" series. Column of 3 mm diameter and 100 mm long, filled by Chromaton, was used at the analysis of plastifiers.

Fluid high pressure chromatography was performed on fluid chromatograph of DuPont Instr. Company in regimes of adsorptional and gel-penetrating chromatography on Zorbax and SE-100 columns.

Plastifier extraction was performed by diethyl ester at 45°C during 6 days.

Molecular-weight distribution of PVC was determined by gel-penetrating chromatography method.

Spectrometer Qualimatic of Digilab Company was used for determination of IR-spectra of the samples.

The analysis of plastifier amount in plasticate were performed according to absorption band in 1720-1726 cm^{-1} range (C=0 valent oscillations) and in 1020 cm^{-1} range (C-O-C valent oscillations).

EXPERIMENTAL RESULTS AND DISCUSSION

Figure 1 shows thermogravimetric curves of PVC-plasticate mass losses, containing various amounts of dioctylphtalate (DOPh), from which it follows, that it is enough to expose PVC samples at 350°C during 25 min in order to obtain reliably the values of residual sample mass after plastifier desorption and dehydrochlorination. In this case mass loss depends linearly on DOPh content at heating up to 400°C.

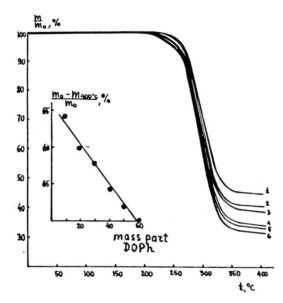

Figure 1. a) Thermogravimetric curves mass losses of CPV-plasticate with different content of dioctylphtalate (DOPh); 10°C/min; 1-10; 220; 3-30; 4-40; 5-50; 6-60 parts per 100 parts of CPV-plasticate. b) Relation between weight loss and content DOPh at heating up to 400°C.

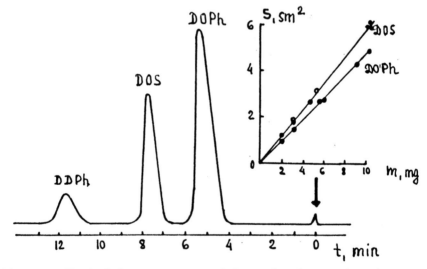

Figure 2. Typical chromotograms of three plastifiers and peak squares versus their amount in specimen; DDPh - disodecilephtalate; DOPh - dioctylphtalate; DOS - dioctylsebacinate.

Figure 2 shows typical chromatograms of three plastifiers and leveling curves, connecting peak squares with plastifier amount in introduced specimen.

Plastifier content in specimen were calculated from spectral curves of UV-absorption according to the expression:

$$C = \frac{Dvl}{\varepsilon m} \cdot 100,$$

where D - optical density at 230 nm for dioctylphtalate in tetrahydrofurane; ε - absorption coefficient; equal (19.3±0.2) l/mg·cm; m - analyzing specimen mass, l - cuvette width.

Concentration of residual plastifier was calculated by density correlation method from IR-spectra of absorption, obtained by ATR method (CRS-5 prism) on films of PVC and plasticate, in order to exclude mistakes, connected with inhomogeneity of sample pressing:

$$\frac{D_f}{D_{tot}} = \frac{X_f \varepsilon_f l_f}{X_{tot} \varepsilon_{tot} l_{tot}} = \frac{X_f}{X_{tot}} Const. \qquad (2)$$

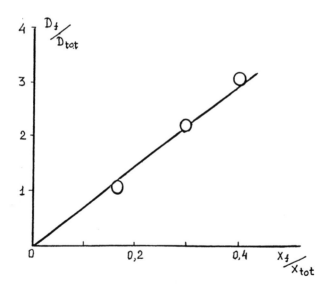

Figure 3. Relative optical density plastifier to polymer versus their relative content.

Figure 3 shows experimental dependence of $\dfrac{C_f}{C_{tot}}$ on $\dfrac{X_f}{X_{tot}}$, where X_f, X_{tot} - concentration of plastifier and polymer in volumeric parts; C_f and C_{tot} - volumeric concentrations of plastifier and polymer.

Experimental data obtained by UV-spectral method, is described satisfactorily by formal-kinetic empyric equation:

$$\ln \frac{m_t}{m_o} = kt^{0.62 \pm 0.33},$$ (3)

where k - desorption constant.

It has been found that chromatographic analysis data are not described by the present equation, because of nonsatisfactory accuracy of the method.

Generalized dependence of diffusional dioctylphtalate desorption from PVC-plasticate, shown on Figure 4 in $\ln \dfrac{m_t}{m_o} - kt^{0.62}$, testifies that the process proceeds, probably, according to one and the same mechanism at all investigated temperatures.

It is natural to expect the change of linear sizes of samples, resulting their aging at diffusional plastifier desorption from PVC-plasticate articles.

Linear and volumeric shrinkages for films were determined according to the formula:

$$\Delta l = l_o - l_t / l_o \text{ or } \Delta V = V_o - V_t / V_o$$

where l_o and V_o - initial parameters; l_t and V_t - parameters after aging during time t.

Shrinkage for wire isolation was determined according to length and diameter of isolation layer

$$\Delta d = d_o - \frac{d_t}{d_o},$$

where d_o and d_t - initial and current parameters, respectively.

It is seen from kinetic curves of mass change and measurements of films and wire PVC isolation, aged thermally in air at 150°C, that general mass change at kinetic curve reaching the plato is (27.0±0.2) % for the film, that is described satisfactorily by (3) equation, and linear size change - by first degree equation:

$$\Delta l = \Delta l_\infty [1 - \exp(1 - kt)]. \qquad (.)$$

The value of rate constant, determined from experimental data, is found $(5.5 \pm 0.2) \cdot 10^{-6}$ s^{-1}.

Linear size change at aging in similar conditions proceeds by lower degree for wire isolation, than for PVC film, that is explained by the presence of adhesion forces of polymer to metal, preventing isolation shrinkage.

Molecular weight distribution (MWD) of initial and thermooxidatively aged PVC-plasticate samples possesses bimodal distribution with peaks, which maxima relate to retaining times of 4.72 ± 0.03 and 7.96 ± 0.02 min, respective values of mean-viscous molecular PVC mass - $4.8 \cdot 10^4$ (A) and $3.6 \cdot 10^3$ (B).

Table 1 shows correlations of PVC MWD components at aging under different temperature conditions. High molecular component content (A) increases monotonously during aging at 140°C. Ag aging during 2.5 hours the peak of low molecular component disappears.

Table 1. Correlation of PVC MWD components for various time - temperature aging regimes.

Aging time, hour	Temperature, °C	PVC MWD component part, %	
		A	B
0	20	70.8	29.2
$4.4 \cdot 10^4$	20	70.6	29.4
100	70	70.5	29.5
50	80	70.4	29.6
100	100	72.5	27.5
100	110	74.9	25.1
100	120	88.0	12.1

At the same time there was observed no sufficient change of MWD (2%), investigating PVC-plasticate samples, exploited in soft conditions during 20-28 years. Probably, sufficient change of MWD at high temperatures in the direction of high molecular component content increase occurs, resulting cross-linking of low molecular PVC fractions.

At the investigation of polyene fragments concentration by fluid chromatigraphy method, forming at dehydrochlorination reaction during aging (0.05% polymer solution in tetrafurane), there appears new peak with retaining time of (1.76 ± 0.01) min, relating to polyene fragments with double bond amount, equal or over five units (retaining time for initial polymer is (2.12 ± 0.01) min.

Table 2 shows parts of peaks with retaining time of (2.12 ± 0.01) (A) and (1.76 ± 0.01) (B) for different conditions of aging.

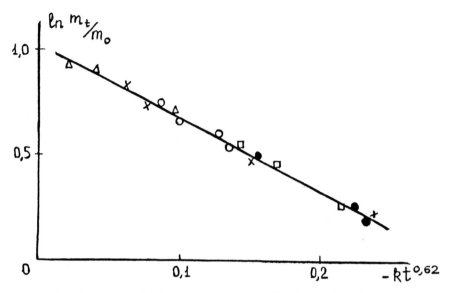

Figure 4. Generalized kinetic curve of diffusional DOPh desorption from PVC-plasticate: - 135°C; x-120°C; -105°C; o-90°C; •-47°C.

Table 2. Correlation of peaks of initial and polyene fragments containing PVC for various time-temperature regimes of aging.

Aging duration months	Temperature, °C	PVC-plasticate mark	Peak part, %	
			A	B
0		I. -40-13	100	-
14	70	-"-	97.2	2.8
305	17	-"-	96.4	3.6
0	70	I. -40-12	100	-
14	17	-"-	97.6	2.4
305		-"-	97.2	2.8
0	70	C-70	100	-
14	17	-"-	97.8	2.2
305		-"-	97.6	2.4

Figure 5 shows initial dehydrochlorination rates in conditional units according to accumulation of polyene fragment sin 120-180°C temperature range. Efficient activation energy, determined from these curves, is 113±5 kJ/mole, which equals to the value of dehydrochlorination energy in presence of the oxygen.

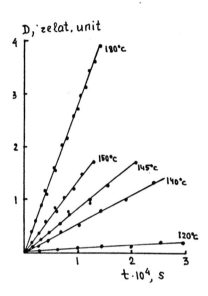

Figure 5. Initial dehydrochlorination rates in conditional units according to accumulation of polyene fragments in 120-180°C temperature range.

Cable isolation, dismantling from articles after 15 and 30 years of exploitation, was investigated in order to clear up the character and physical and chemical process intensities, occurred at low temperature dark aging of PVC-plasticate.

Exploitation was performed in conditions of darkness at equivalent temperature of ~20°C. Minimum temperature was -20°C, minimum one reached +28°C. The main investigation results are shown in Table 3.

Table 3.

Exploitation time, years	Dioctylphtalate amount, mass%	Shrinkage, %	Elongation at break %	Tg, °C	Pore volume cm³/g
0	35±1	-	600±40	-40	0.017
15	27±2	5±2	480±60	-20	0.050
30	15±6	15±6	300±80	-20	0.065

There was observed no sufficient amount of products of thermoxidation degradation and dehydrochlorination by methods of IR-spectroscopy and fluid chromatography. MWD was not recorded.

One can testify, basing on the data obtaining, that diffusional desorption of plastifier is predominant process at low temperature PVC-
plasticate aging. In this case the proceeding of this process may be
recorded by two indexes accurately enough: the decrease of plastifier
amount in the material and the decrease of pore volume in it (mercuric
porometry method).

Studying light aging of PVC, it was started from the supposition,
that aging occurs in surface layer and depends on spectrum light composition. PVC films of various thickness, obtained at slow evaporation of
solvent (dichlorethan) from 5% polymer solution, were irradiated by
light with λ=254 nm (bacterial lamp DB-60) and λ>300 nm (DB-120).
PVC oxidation was controlled by accumulation of carbonile groups, determined by IR-spectroscopy method according to optical density at
1720 cm^{-1}, dehydrochlorination - by polyene formation, determined according to absorption spectra of films in UV- and visible light range.

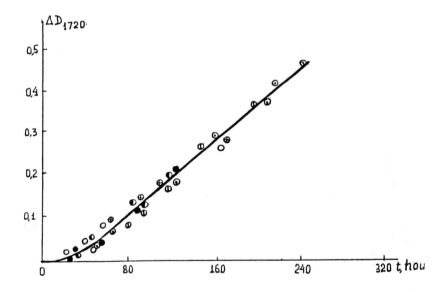

Figure 6. Kinetic curves of accumulation of carbonile groups at light
irradiation with λ=254 nm of the films different thickness.

Figure 6 shows kinetic curves of accumulation of carbonile groups and polyenes at light irradiation with λ=254 nm. It follows from the data, that oxidation degree does not depend on sample thickness, which testifies the proceeding of photooxidation processes in PVC in very thin surface layer, possessing thickness less than 20 μm, whereas dehydrochlorination degree depends strongly on sample thickness in consequence of degradation process proceeding in the volume.

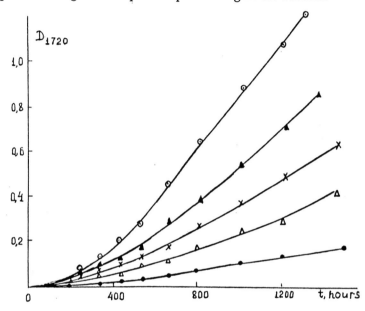

Figure 7. Kinetic curves of accumulation of carbonyle groups at light irradiation with λ>300 nm of PVC films different thickness on air at ambient temperature: 1-25 μm; 2-37μm; 3-53μm; 4-75μm; 5-105μm.

Kinetic curves of PVC aging were found principally different under the influence of softer UV-radiation with λ>300 nm (Figure 7): both oxidation and dehydrochlotination degrees depend sufficiently on sample thickness, but in this case aging proceeds irregularly in the bulk, because kinetic curves do not transform into general dependence on D/*l*-t coordinates (D - optical density, *l* - film thickness, t - radiation time). Consequently, in these conditions oxidation proceeds preferably in deeply dislocated polymer layers, and not in the surface one. Differences in dehydrochlorination kinetics at light irradiation by λ>300 nm and λ=254 nm show not qualitative character, but quantitative one only an din absence of clearly displayed induction period in the case of radiation by λ>300 nm. One more important feature of PVC was found experimentally. It is the dependence of absorption spectrum of the sample on spectral composition of influencing light (Figure 8), that is connected, apparently, with the differences in the composition of forming polyenes. Sensitivity of PVC samples to spectral light composition can be used in practice for the determination of the place of sample aging, because spectral composition of sun light on Earth surface depends on latitude of test place and ozone layer thickness.

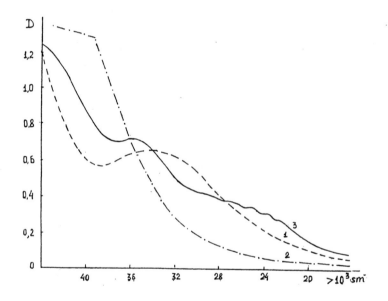

Figure 8. Absorption spectrum of the PVC films after light irradiation with λ=254nm (1); λ>300nm (2) and total light Hg lamp high pressure irradiation (3).

Next desorbing component of PVC-plasticate is stabilizer, which can diffuse to sample surface at long contact with water and solve in it. It was investigated the desorption of lead salts from PVC-plasticate films, containing lead sulphate and stearate, at 20, 45 and 70°C during 10 days. Salt content was determined on atomic-adsorptional spectrophotometer. The curve of lead salt desorption (Figure 9) possesses two characteristic parts: the part of fast desorption - exponential one; linear part - slow extraction.

Exponential part of desorption is described satisfactorily by Fick's equation. Diffusion coefficients, shown in the Table, were calculated from kinetic curves.

It should be mentioned, that diffusion coefficient of stabilizer in experimental temperature range depends insufficiently on it. Moreover, stabilizers possess different mobility in the material, saturated by water.

Contact of PVC-plasticate films with dropwise water leads to relatively fast washing out (during 10 days) of easily desorbing part of lead salts.

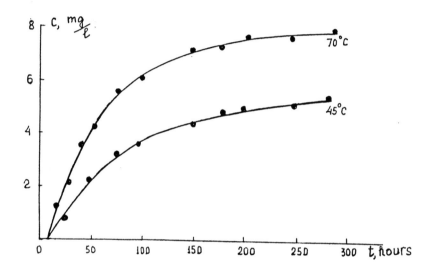

Figure 9. Desorption of lead salt from PVC-plasticate in water.

Thus, the conduction of control tests on lead salts desorption from unknown PVC-plasticate into water may characterize conditions of previous exploitation or exposure: for example, fast extraction of 5-7% lead salts may testify material exploitation in dry climate, and absence of such desorption - exploitation in humid climate.

The samples were investigated in analog to PVC ones, tested in different aging conditions, which were exploited on health resort in open air in Batumi (hot) and Jakutsk (cold) regions during 3 years (1987-1990).

Table 4. Coefficients of lead salts diffusion in PVC-plasticate (30 mass% of dioctylphtalate) at different temperatures.

Stabilizer content, mass%	Film thickness, cm	T, °C	$D \cdot 10^9$, cm^2/s
Lead sulphate 0.7	0.10	20	0.59
		45	0.69
Lead stearate 0.4		70	0.86
Lead sulphate 2.0	0.20	45	6.10
		70	6.30
Lead sulphate 1.6	0.30	20	7.20
		45	7.70
		70	8.30
Lead sulphate 1.5	0.35	20	5.00
		45	8.30
Lead stearate 2.5		70	11.00

Testing the samples, one based on the fact that mostly characteristic parameters for determination of aging degree may be mechanical indexes, plastifier amount, pored structure change. The data are shown in the Table. Naturally, the most informative parameter is plastifier amount, which decreases sufficiently during PVC-plasticate aging.

Table 5. The change of plastifier content and mechanical properties of PVC-plasticate films during aging in different climate zones.

Material	Aging conditions	σ_b, MPa	σ_b, %	C_{DOP}, mass%
CPV-DOP	Initial	39±3	280±30	17±1.0
(17 mass %)	Jakutsk			
	6 months	35±6	180±100	15±1.0
	12 months	23±2	44±30	10±0.5
	24 months	27±2	42±35	10±0.5
	36 months	32±2	23±20	10±0.7
	Batumi			
	6 months	34±5	170±80	15±1.0
	12 months	25±2	60±20	12±1.0
	24 months	28±2	40±25	10±0.7
	36 months	30±2	25±10	10±0.7
CPV-DOP	Initial	32±2	400±40	23±1.0
(23 mass %)	Jakutsk			
	6 months	28±4	240±40	21±1.0
	12 months	22±2	260±80	17±1.0
	24 months	20±2	56±27	10±0.7
	Batumi			
	6 months	27±3	280±50	20±1.0
	12 months	23±2	220±40	16±1.0
	24 months	20±2	100±30	11±1.0

Thus, the main indications of aging of PVC-plasticate samples from film materials, aged in model conditions, climatic conditions, and occurred for a long time under exploitation, were established with the help of various methods of separation of sample microamounts and the application of complex of modern physicochemical analysis methodics, and mechanism of their factors was determined, also. It was shown, that losses of exploitation and other functional properties of PVC-plasticate occur, first of all, resulting diffusional desorption of plastifier and other additives.

One can form an opinion on conditions of material aging with the help of experimentally deduced empyric kinetic equations of mass loss or additive spending during exploitation or exposure.

The results, been obtained point out also improper application of any general macrokinetic prognosing equation for description of the proceeding of PVC-plasticate degradation at unknown climatic or exploitation aging.

REFERENCES

1. N.M. Emanuel, *Proceedings of Lectures IY Polymer School*, Tallin, 1970.
2. K.S. Minsker, *Intern. J. Polymeric Mater.*, 24, 235 (1994).
3. K.S. Minsker, S.V. Kolesov, G.E. Zaikov. *Degradation and Stabilization of Polymers*, Utrecht, VSP Science Press, 1988.
4. H.T. Bowlly, D.G. Gerrard, K.S.P. Williams, I.S. Biggin, *J. Vinyl Technol.*, 8, 176 (1986).
5. F. Soundhelmer, D.A. Ben-Efrain, R.J. Wolowsky, *J. Amer. Chem. Soc.*, 183, 1675 (1961).

THERMOSTABILIZATION OF POLYISOPRENE BY PHENOL GRAFTING IN CARBENE SYNTHESIS

E.Ya. Davydov, V.P. Pustoshnyi and A.P. Vorotnikov
Institute of Chemical Physics RAN, 4 Kosygin str.,
Moscow 117977, Russia

The reaction of 2,6-di-tert-butylcyclohexadienone carbenes with polyisoprene has been used to graft 2,6-di-tert-butylphenol on macro-molecules by thermolysis of quinonediazide. The oxidation of polyiso-prene with grafted phenol and analogous low molecular antioxidant was studied at 353K ad 363K by yields of carbonyl and hydroxyl groups and by the oxygen sorption. The induction period of oxidation is rised linearly for grafted antioxidant by its concentration increase within the range of $6 \cdot 10^{-3}$-$4 \cdot 10^{-2}$M. The consumption of grafted phenol occurs as a result of linear terminations of oxidation chains.

The limitation of traditional antioxidant applications in polymers is connected often with their physical losses owing to the volatiliza-tion and the washing-off by water and other organic solvents. The ef-fective method to avoid losses of stabilizers is the binding with macromolecules by chemical reactions. The grafted antioxidant can be effective under service conditions of polymeric materials at high tem-peratures and also their reprocessings.

The termination of oxidation chains on antioxidants chemically linked with macromolecules is realized under conditions of high molec-ular mobility by physical diffusion of free valence. It is reasonable in this context to use the grafting for elastomer stabilization [1,2]. The stabilization effect can be achieved for these flexible polymers in whole temperature range of use and reprocessing. The termination of kinetic chains is limited on the other hand by chemical diffusion of free valence in rigid polymers below the glass temperature with the result that grafted stabilizers are become uneffective.

The aromatic nitrozocompounds are used most commonly as rubber modifying agents which give products similar by structure to N-phenyl-p-phenylenediamine [3,4]. However, the graft method of such antioxidants has some limitations. This process is continued 10-20 hours even at 370K. The conversion degree does not exceed 50-70% due to side reactions [3]. The secondary products of nitrozocompound conversion can produce the coloring of rubbers at vulcanization. Besides the modification is followed by marked destroying of macromolecules. The application of nitrozophenols instead of amine analogs allows the coloring of goods to be avoided. But they give weaker grafted antioxidants.

The convenient method of the stabilizer fixation in polymers is the reaction of suitable carbenes with macromolecules. There are examples of effective use of carbenes to stabilize polyolephynes by 2,6-di-tert-butylphenol (DTBP). The fixed phenol indicates good protective properties at 300-500K exceeding those of ungraftedphenols [5]. The similar method was used to fix DTBP on polyisoprene (PI) in this work. The comparative estimation of antioxidational activity of grafted and low-molecular phenols has been made.

EXPERIMENTAL

PI films with 2,6-di-tert-butylquinonediazide (DTBQA) additives ($6 \cdot 10^{-3}$-$4 \cdot 10^{-2}$M) were heated in the course of one hour at 343K to graft DTBP. In this case the appearance of 280 nm phenol band in UV spectrum is observed. The reprecipitation of PI samples does not lead to the band disappearance. The estimation by value $\varepsilon_{280}=1.7 \cdot 10^3$l mol^{-1} cm^{-1} shows that the phenol concentration is about 90% of that of DTBQA inserted to PI. The oxidation of PI with grafted DTBP at 353K and 363K was performed on 50µ films placed on fluorite plates. The thermooxidation kinetics was determined by the accumulation of carbonyl and hydroxyl groups and by the oxygen sorption. The formation of C=O and OH groups was followed by 1720 cm^{-1} and 3200-3600 cm^{-1} bands respectively. The oxygen sorption kinetics was studied by air pressure changes in manometric unit. The oxidation of PI stabilized by 2,6-di-tert-butyl-4-methylpropionatephenol (DTBMP) was studied by the same methods to compare both antioxidants from the point of view their stabilizing capabilities.

RESULTS AND DISCUSSION

The formation of phenol derivatives in carbene synthesis is possible in PI by reactions of triplet carbenes with C-H allyl bonds and as a result of spirodienone conversion [6]

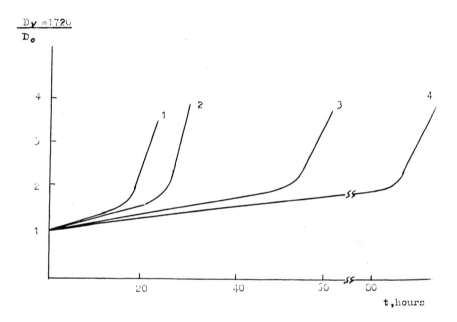

(1)

The first mechanism is typical of quinonediazides photolysing at low temperatures. The second reaction is principal at thermolysis.

Figure 1. The kinetics of C=O accumulation at 353K in PI stabilized by DTBP grafting: $[IH]_0 = 6.36 \cdot 10^{-3}M$ (1), $1.27 \cdot 10^{-2}M$ (2), $2.12 \cdot 10^{-2}M$ (3), $4.24 \cdot 10^{-2}M$ (4).

The kinetic curves of carbonyl group accumulation in PI with DTBP are shown on Figure 1. The same curves were obtained for hydroxyl products and oxygen sorption. The dependences of induction periods of oxidation on grafted antioxidant concentrations are shown in Figure 2. The analogous relationships are presented for low-molecular phenol. It is clear that observed concentration dependences have essential differences. The induction period is rised linearly for grafted antioxidant at the increase of its concentration. The evident departure from linearity was observed for DTBMP within the concentration range up to $3.8 \cdot 10^{-2}$M. The induction period values are close for both antioxidants at low concentrations. But the difference of τ rises largely with antioxidant concentration increase. This result allows to consider that the deviation from linearity of the concentration dependence of τ for DTBMP does not result from its sweating-out that is typical of low-molecular antioxidants.

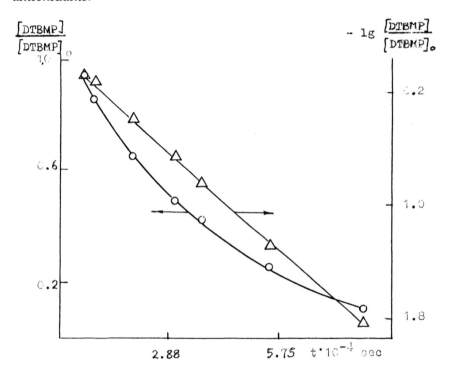

Figure 2. The dependence of the induction period of C=O formation on antioxidant concentrations in PI at 353K: 1 - grafted DTBP, 2 - DTBMP.

It is known that the dependence of τ on the inhibitor concentration at thermooxidation is defined by kinetics of the antioxidant consumption. Thus PhOH consumption with the constant rate corresponds to the linear concentration dependence of τ [7]. Then the low rate limit of phenol consumption is achieved determining by initiation rate of the oxidation $W_0/2$. The kinetics falled off from the zero order not until at the end of the induction period when phenol concentration is approached to a critical value. Yet neither proportionality of induction period values and phenol concentrations nor the zero order of its consumption kinetics are not accomplished commonly in practice.

The origin of this effect is consisted in unexpected side reactions with the assistance of phenoxyl radicals, forming by the termination of oxidation chains [8], as well as direct interaction of phenols with oxygen in the course of inhibited oxidation [9]. The rate of the inhibitor decay within the induction period is rised in consequence of side reactions and the kinetics of the process is followed as a rule to the first order law. Thus the dependence of τ on [PhOH] is linearized in half logarithmic coordinates at concentrations exceeding the critical value [10]. Such dependence is specific for the oxidation of PI with DTBMP (Figure 3). The critical concentration of this antioxidant is $2 \cdot 10^{-3}$M at 353K.

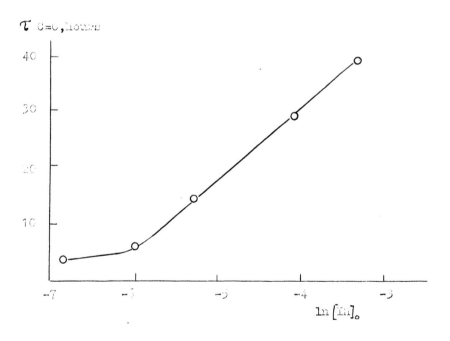

Figure 3. The dependence of the induction period of PI oxidation on logarithm of DTBMP concentration at 353K.

One can regard possible reasons resulting in the decrease of DTBMP antioxidational activity in PI as compared with grafted DTBP. The direct oxidation of low-molecular phenol does not take place apparently for lack of this process for its grafted analog. Some side reactions of phenoxyl radicals can be expected from the same considerations such as reactions with hydroperoxides, C-H bonds of macromolecules and likewise reactions of quinolide peroxide formation and conversion [11]. But the possibility of p-position can be assumed with the result that slightly active inhibitor radicals is converted to active substitute radicals

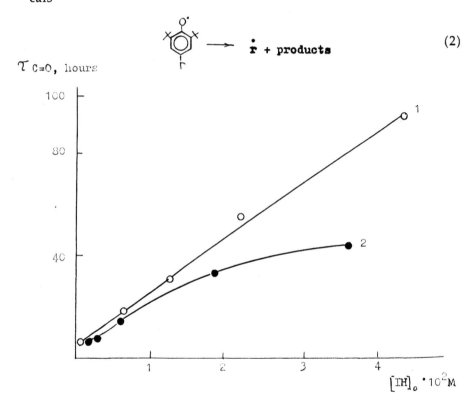

Figure 4. The kinetics of DTBMP consumption in PI at 353K.

The stationary concentrations of peroxide radicals, hydroperoxides and the phenol consumption rate are increased in consequence of the reaction (2). The similar reactions are typical of p-alkoxiderivatives of phenols [12]. Then PhOH consumption is accelerated by far within a range of phenol high concentrations. The zero kinetic order is substituted by first order in limit. The reaction (2) can be ignored for phenoxyl radicals containing alkyl substitutes in p-position [12]. So grafted DTBP is not affected by undesirable β-dissociation decreasing its efficiency.

The Figure 4 shows that DTBMP consumption is described fairly by the first order of law. One can assume on the base of this fact that DTBMP conversion takes place likely to (2)

$$\cdot O-\bigcirc^{X}_{X}-CH_2CH_2OCOCH_3 \longrightarrow O=\bigcirc^{X}_{X}=CH_2 + CH_2O + CO + \dot{C}H_3 \qquad (3)$$

The comparison of antioxidational activity of traditional and grafted antioxidants shows that the last has qualitative advantages in addition to the non-volatility. The grafted DTBP is "ideal" antioxidant in overall studied concentration range the upper limit of which corresponds to concentrations used in real working conditions. Its consumption within the induction period occurs as a result of linear terminations of oxidation chains and is not attended evidently by side reactions. These advantages permit to make use of space-hindered phenols grafted by carbene synthesis on macromolecules to stabilize PI and other elastomers.

LITERATURE

1. K.B. Piotrovskii, Z.N. Tarasova. *Aging and stabilization of synthetic rubbers and vulcanizates.* Khimiya, Moscow, 1980, p. 237 (in Russian).

2. D. Barnard, P. Lewis, *Natural rubber science and technology,* ed. A.D. Roberts, 1988, Oxford University press, Oxford, v. 2, p. 130.

3. L.M. Kogan, B.A. Krol, L.M. Davydova, M.B. Monasyrskaya, O.I. Bel'govskaya. *Vysokomolek. Soed.,* 1976, v. A18, N5, p. 1076-1081 (in Russian).

4. S.M. Kavun, T.V. Fedorova, G.I. Akin'shina, .Z.N. Tarasova, I.V. Hodjaeva, *Vysokomolek. Soed.,* 1973, v. A15, N10, p. 2378-2381 (in Russian).

5. M.L. Kaplan, P.G. Kelleher, G.H. Beggington, R.L. Hartless, *J. Polymer Sci.,* Polymer Lett. Ed., 1973, v. 11, N6, p. 357-361.

6. V.V. Ershov, G.A. Nikiforov, A.A. Volod'kin, *Space-hindered phenols,* Khimiya, Moscow, 1972, p. 204 (in Russian).

7. V.A. Roginskii, *Kinetika i kataliz,* 1982, v. 23, N5, p. 1081-1088 (in Russian).

8. V.A. Roginskii, *Vysokomolek. Soed.*, 1982, v. 124, N9, p. 1808-1827 (in Russian).
9. E.L. Shanina, V.A. Roginskii, G.E. Zaikov, *Vysokomolek. Soed,* 1986, v. B28, 9, p. 1971-1976 (in Russian).
10. Yu.A. Shlapnikov, *Uspehi khimii*, 1981, V. 50, N6, p. 1105-1140 (in Russian).
11. V.V. Pohelintcev, E.T. Denisov, *Vysokomolek. Soed.*, 1985, v. 27, N6, p. 1123-1136 (in Russian).
12. V.A. Roginskii, V.Z. Dubinskii, V.B. Miller, Izv. An SSSR, ser. khimicheskaya, 1981, N12, p. 2808-2812 (in Russian).

ESTIMATION OF HEAT AND RADIATION STABILITY IN ELASTOMERS

N.R. Prokopchuk, A.G. Alekseev, T.V. Starostina and L.O. Kisel

Institute of Physical and Organic Chemistry
Belorussian Academy of Sciences, Minsk
Russian Research Center for Elastomeric Materials and Wares,
S. Petersburg

Elastomers have to be of high stability (durability, τ) because of their long-term operation as parts of articles under strict conditions, most important of which are high temperatures and radiation. The GOST (State Standard) methods of determining τ, widely practiced and based on the heat aging of elastomers, are time- and material-consuming. They are unfit for high temperatures and make it impossible to estimate the contribution of different outer factors into decrease of rubber durability. A method developed by the present authors [1] lacks those drawbacks. It permits determination the durability of elastomers 25-30 times faster, consumes a small quantity of material (a 120x140x1 mm rubber plate), makes it possible to extend a temperature range of τ determination and estimate the contribution of ionizing radiation into its decrease. It is a feature of the new method allowing thermo-oxidizing destruction to be performed at the temperatures similar to those used in the GOST methods but requiring an additional superposition of the field of tensile stress which accelerates the destructive processes in samples and considerably decreases the time required for estimation of activation energy, U_o, and, consequently, determination of material durability. The method involves a common dependence (see the figure) binding the rubber durability and activation energy of their thermo-oxidizing destruction, E, found from the extensive experimental material on rubber durability by means that of the Gost method based on continuous thermal aging. It also includes an express-procedure for U_o estimation according to the temperature dependence of ultimate tensile stress when thermo-oxidizing destruction is accelerated by the superimposed field of mechanical forces.

Experimental dependence of durability in air at 25°C on the activation energy of thermal aging for rubbers on diferent polymeric bases and composition.

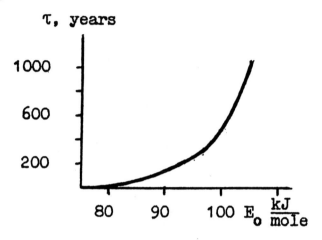

In replicate tests by using method [1] and GOST 9713-86, the activation energis of processes and durability of 20 rubbers have been determined. The U_0 and E turned out to coincide in all the cases to an accuracy of ±2 kJ/mole while the durability values correlate with one another. Therefore, the durability of elastomers can be determined according to the empirical formula:

$$\tau = 10^{(\alpha U_o + \beta)} \cdot \exp(U_o / RT),$$

where α=-0.111±0.001; β=-3.687±0.05.

The essence of the express-procedure for U_0 determination consists in stretching performed at temperatures of the corresponding interrelaxation interval, at loading rates and clamping length ensuring the linearity of tension diagrams and destruction in 100-400 s. In this case parameter U_0 is determined by the formula:

$$U_o = T_o R \ln(\tau_{eff} / \tau_o),$$

where T_0 is the temperature determined by linear extrpolation of the temperaure dependence of ultimate tensile stress $\sigma(T)$ within an inter relaxation interval, σ=0;

τ_0 - a constant equal to 10^{-12} s for rubbers;

τ_{eff} - the effective durability at indicated test conditions on a tensile-testing machine, calculated by the formula:

$$\tau_{eff} = 3.7 \cdot 10^{-2} \frac{T_m}{T_o - T_m}$$

where T_m is the temperature of an interval middle within which U_o is determined.

To demonstrate the new method possibilities, the rubber durability was determined on the basis of ethylen-propylene rubber in a wide temperature range, including higher temperatures, when the GOST methods do not work. The tests were performed on a yMUB-4 device. The samples had the form of a double blade with a 10x1x1 m operating part, tension rate of 5 mm/min; heating time of 8 min; temperature range of 20-300°C.

Table 1 presents the initial data to calcultae U_o, as well as the computed values of U_o and τ corresponding to three temperature intervals.

Table 1. Calculation of rubber durability based on the ethylene-propylene raw rubber

Stage	Temperature range, °C	Data for calculation	U_o kJ/mole, τ
I	20-60	$T_o=86$°C $\varepsilon_p=80\%$ t=96s $\tau_{eff}=22$s	$U_o=90$ $\tau25°=125$ years
II	60-140	$T_o=246$°C $\varepsilon_p=20\%$ t=24s $\tau_{eff}=2.2$s	$U_o=122$ $\tau140°=5.1$ days
III	250-300	$T_o=300$°C $\varepsilon_p=15\%$ t=18s $\tau_{eff}=10$s	$U_o=143$ $\tau300°=8.7$ sec.

In every temperature interval a relative elongation at rupture, ε_p, was practically constant at all the temperatures within an interval which indicated a stable structure inside them. The transition from low to higher temperature intervals is accompanied by increase in U_o which can be explained by partial linkage (formation of a three-dimensional net) in rubber.

This is also evidenced by a drastic decrease in ε_p from 80% to 15%. Comparison of the U_0 values obtained by the GOST method on the thermal aging data (70, 90, 110, 195 and 150°C) and by the developed express-procedure in the second temperature range (60-140°C) shows a good convergence of results: 126 and 122 kJ/mole, respectively. The determination of U_0 by the GOST method within the I and III intervals is impossible. Therefore, the calculation here was performed by the express-procedure. The $\tau_{300°}$ value obtained corresponds to the data of bench tests.

Besides, the possibility of estimation of the ionizing radiation contribution into a decrease in durability has been shown exemplified by rubbers from butadiene-nitrile (CKH-18CM) and chlorprene (nairite M) raw rubbers. To this end a half of samples of two rubbers was ^{60}Co-γ irradiated with a dose of 50 Mrad (500 kgRa), a dose rate of 400 rad/s (4 gRa/s). As it follows from Table 2 the ultimate tensile stress of CKH-18CM samples under irradiation decreases 2.1 fold and that of nairite-M samples, 3.4 fold, while parameter U_0, by 7 and 16 kJ/mole, respectively.

Table 2. Variation of rubber particles under ionizing irradiation [2].

Sample	Initial			After irradiation		
	σ,MPa	U_0, kJ/mole	τ^*, years	σ, MPa	U_0, kJ/mole	τ^*, years
CKH-18CM	30	109	1894	14.2	102	677
Nairite-M	75	121	11033	22.2	105	1063

*Durability correspond to 25°C.

The effect of γ-radiation on durability of rubbers on different polymeric base is inadequate: τ of irradiated CKH 18CM samples amounts to 35% of the initial samples while for nairite-M altogether 9.6%. At the same time, the durability of both rubbers after 500 kgRa irradiation is sufficiently high to satisfy specifications to rubbers stocking different technical objects [2].

Thus, the results of the present study show that the express-procedure developed has a number of advantages against the known methods used for estimation of durability of polymeric materials.

REFERENCES

1. Prokopchuk, N.R., Alekseev A.G., Starostina T.V., Kisel L.O. *Method for determination of rubber durability,* Dokl. AN BSR. 1990 v. 34, N11, p. 1026-1028.
2. Alekseev A.G., Starostina T.V., Kisel L.O., Barchenko S.V., Kudryashova T.A., Prokopchuk N.R. *Effect of ionizing radiation on the durability of elastomeric materials.* Izv. ANB. Ser. Khim Nauk. 1992 N2, p. 108-110.

Stabilization of Cationic Acrylamide Polymers Designed to Enhance Oil Recovery

I.A. Golubeva, V.F. Gromov, Ye.V. Bune,
O.N. Tkachenko, and R.H. Almaev

To enhance oil recovery, water-soluble acrylic polymers have been lately increasingly used. Their principal designation is to compensate for heterogeneity of producing formation and to extend flooding sweep area. That can be achieved by means of water thickening, by addition a certain amount of polymers in the process of conventional flooding of oil-bearing formation, as well as by limiting water influx by producing a plugging agent owing to focculating properties of polymers. These polymer properties can be utilized in solving ecological problems - in treatment of oil-containing waste water.

The authors have studied possible means to synthesize and stabilize high-molecular water-soluble cationic acrylamide (HA) polymers designed to enhance oil recover processes.

Synthesis of cationic polymers was carried out by means of radical copolymerization of acrylamide with dimethyldiallylammonium chloride (DMDAAC) or dimethylaminoethylmethacrylate sulfate (AM.HA) under isothermic conditions or at adiabatic heating-up of the reaction mixture. Adiabatic mode of polymerization process was chosen because it most closely corresponded to industrial production conditions. It was difficult to maintain constant temperature in large reactors during synthesis of polymers due to substantial heat evaluation and high viscosity of reaction mixture which developed in these systems already on early process stages.

Technical solution of DMDAAC of specific concentration (~40% by weight) was used in the study.

It was found that addition of acrylamide to DMDAAC caused us botanical increase in polymer molecular weight (Table 1). High-molecular copolymers acrylamide with DMDAAC were produced both in isothermic and adiabatic regimes. Variations in initiator concentration and reaction temperature just insignificantly affected molecular weight and properties of polymers, apparently, because of determinative influence of reaction of chain transfer through DMDAAC molecules on molecular weight of these polymers.

Table 1. Copolymerization of acrylamide with DMDAAC.

Polymerization conditions	AA/DMDAAC mol. %	[M] mol/l	[PK] mol/l	$T_{init.}$ °C	$[\eta]$ dl/g
Isothermal	30/70	3,7	1.10^{-2}	30	2,9
	30/70	3,7	1.10^{-2}	60	1,7
	0/100	2,6	1.10^{-2}	60	0,44
	30/70	3,7	1.10^{-3}	30	2,9
	30/70	3,7	1.10^{-3}	60	2,1
	0/100	2,6	1.10^{-3}	30	0,39
	0/100	2,6	1.10^{-3}	60	0,49
Adiabatic	48/52	4,3	1.10^{-3}	25	4,1
	48/52	4,3	$0,8.10^{-3}$	25	3,7
	48/52	4,3	$2,9.10^{-3}$	45	3,7
	16/84	3,0	$7,9.10^{-3}$	25	1,9
			$2,9.10^{-3}$		

Copolymerization of acrylamide with AM.SA was studied in a wide range of monomeric mixture compositions, at different total concentration of monomers, initiator and chain transfer agents. The reaction was carried out in presence of potassium persulfate, at initial temperature 25-30°C.

It is known that in order to produce high-molecular polymers the reaction should be performed in concentrated solutions. However, in the process copolymerization of acrylamide with AM.SA in adiabatic conditions, variation of total concentration of monomers in range of 1,5 to 4,5 mil/l has little effect on molecular weight of produced copolymers and their properties. Measured by light scattering method the values of M_W for copolymer samples which were produced at monomers concentration 1,4-4,5 mil/l was practically constant and equaled 3.10^6 (Table 2).

As we may suppose, the unusual dependence of polymer molecular weight upon monomers concentration is a specific peculiarity of radical adiabatic polymerization. In these conditions molecular weight of the polymer is subject to such contrary factors as (1) increase of monomers concentration which would elevate molecular weight of the polymer, and (2) increase of maximum temperature of reaction mixture which is more pronounced at high monomers concentrations and provokes increase in rate of initiation and chain transfer reactions, and, consequently, causes decrease in molecular weight of the polymer.

Table 2. Effect of total monomer concentration on molecular weight of AA-DM.SA copolymers (AA/DM.SA=60/40, [PK]=2,5.10^{-3} mol/l, initial temperature 35°C)

[M] mol/l	[η] dl/g	$M_w.10^{-6}$
4,5	4,6	3,6
3	4,5	3,0
2	4,1	2,6
1,5	5,1	3,5

Table 3. Effect of monomeric mixture composition on molecular properties of AA-DM.SA copolymers ([Butanol]=0.1 mol/l, [PK]=1,0.10^{-3} mol/l, T$_{init}$=25°C).

AA/DM.SA	[M] mol/l	[η] dl/g	$M_w.10^{-6}$
90/10	3,0	15,0	6,4
80/20	4,5	11,8	5,3
60/40	4,0	12,8	5,8
30/70	4,0	5,4	6,3
70/30[a]	3,0		4,8
30/70[a]	3,0		5,6
0/100[a]	3,0		4,3

When increasing acrylamide content in monomeric mixture, increase in intrinsic viscosity of copolymers is observed (Table 3). Taking into consideration the fact that molecular weight of these polymers is practically the same, it may seem that this effect is associated with structural difference in macromolecules produced from different monomers mixture compositions.

Insignificant dependence of copolymer molecular weight upon monomeric mixture composition during copolymerization of acrylamide with AM.SA correlate with kinetics data of acrylamide copolymerization with dimethylaminoethylmethacrylate salts in diluted aqueous solutions. Indeed, the rate of acrylamide homopolymeriation is higher than that of second commoner. However, for monomeric mixture composition AA: aminoether salt in the range from 80:20 to 30:70, the total rate of copolymerization is approximately constant and close to the rate of aminoether salt homopolymerization.

In some cases, during the process of copolymerization of monomeric mixtures with high acrylamide content, and, especially, when processing such polymers into powder, insoluble in water polymers are produced. Loss of solubility in such a case may be determined both by a physical factor (particularly, by entanglement of high-molecular chains resulting in extremely slow dissolution rate), and by chemical cross-linking of the polymer under the impact of mechanical stress in the process grinding. Insoluble polymer generation during processing may as well be affected by monomer and/or initiator traces retained in the powdery polymer.

To produce completely soluble products in the process of copolymerization, the reaction was performed in presence of a chain-transfer agent; butane and glycerin were used for this purpose. Introduction of such additives in the reaction mixture provided for lower molecular weight polymer generation with decreased branching and, consequently, with superior solubility. Besides, decreased branching of polymer chains provided for their higher stability. Addition of 0,16 mil/l of glycerin to reaction mixture produced soluble copolymer with intrinsic viscosity 10 dl/g. To produce a completely soluble polymer in butanol presence, its concentration should be at least 0,5 mol/l.

To prevent polymer cross-linking in its grinding process, a number of stabilizing agents was studied, comprising: phenoxane (potassium salt of phenozanic acid - PhA), phenozane (composition of phenozanic acid and potassium carbonate - PhZ), mercaptobenzimidazole (MBI), sodium dimethyldithiocarbamate (DDC), thiourea (ThU) and phosphirc acid. Stabilizing effect of PhA and PhZ is based on their with interaction with peroxide radicals $RO_2^{\cdot}$, effect of MBI is based on destruction of hydroperoxides, while DDC displays a complex effect, in particular, destruct hydroperoxides and interacts with peroxide radicals.

Stabilizing additives were introduced into the polymer before its grinding. Amount of the introduced additives equaled 0,3-10% (to the polymer) and was determined by its water solubility.

It was found that PhA and PhZ not only failed to improve polymer solubility but also their introduction into the polymer during processing stage was accompanied by cross-linking of polymer chains. In presence of DDC polymer solubility improved, however, this effect was possibly associated with certain decrease in molecular weight. Substantial stabilizing activity is peculiar to MBI and ThU. Polymers do not lose their solubility during processing in presence of these compounds. As a rule, intrinsic viscosity of such polymers is higher than that of polymer processed in absence of stabilizing agents. Higher as well were the screen-factor values (used to characterize viscoelastic properties of aqueous polymer solutions) of the polymers which had been processed in presence of MI or ThU (Table 4). Besides, in some cases, introduction of stabilizing agent prevented formation of the insoluble polymer during its processing into powder.

To ascertain the possible synergistic effect in mixtures of individual stabilizers, the system MBI + PhA was studied. The obtained results revealed that concerning its efficiency, mixture of these stabilizers did not differ much from stabilizing effect of its individual components.

Table 4. Effect of stabilizing agents on viscoelastic properties of aqueous polymer solutions (polymer concentration 0.1%)

AA/DM.SA monomeric mixture composition, mol.%	Powdered polymer properties								
	Without stabilizing agent		MBI 3%		PhZ 1,5%		DDC 10%		ThU
	[η]	SF	[η]	SF	[η]	SF	[η]	SF	[η]	SF
80/20	9,3	18	10,1	32	-	-	4,9	-	12,2	31
80/20	7,4	18	8,0	22	3,6	-	2,5	-	-	-
80/20	5,6	7	6,5	9	h.p.	-	-	-	-	-
60/40	4,3	6	6,3	9	4,7	8	3,7	-	-	-
60/40	2,5	2	9,2	19	-	6	1,0	-	-	-

Investigation the effect of stabilizers on viscous and viscoelastic properties of aqueous polymer solution, copolymers of acrylamide with DM.SA of different compositions were used. Analysis of obtained results revealed that the found relationships were valid for all investigated samples. Acrylamide content in these samples varied in the range of 80 to 30 mol %.

Cationic polymers produced in this work and processed in presence of stabilizers can be recognized as capable to improve significantly solution viscoelastic properties and in this respect they may be practically used as water thickeners, e.g., in oil production processes. Synthesized products are peculiar for relatively high screen-factors values.

Flocculation properties of synthesized stabilized polymers were also studied. Flocculation effect was estimated by bentonite suspension in mineralized water settling rate. It was shown that unlike the completely cationic industrial polymer WPC-402 which did not affect settling rate, addition of just small amounts of AA-DM.SA copolymer to suspension was accompanied by a sharp increase in flocculation rate. As to their efficiency in such systems, cationic AA-DM.SA copolymers containing 60-80 mol.% of acrylamide moieties and with molecular weight about 7.10^6 surpassed as well acrylamide homopolymer of the same molecular weight.

Thus, techniques to produce completely water-soluble powdered cationic acrylamide polymers of high operational performance were found in this study.

MECHANOOXIDATIVE DEGRADATION AND DESTRUCTIVE KINETICS OF ELASTOMERS

D. Saidov, J.T. Achrorov and G.E. Zaikov*

Samarkanol-city, Samarkand State University, Uzbekistan
**Institute of Chemical Physics, Moscow, Russia*

Mechanisms and the kinetics of mechanooxidative degradation of elastomers are considered. The role of the oxidative degradation in the durability of elastomers is shown. The aspects of both the mechanooxidative degradation in the absence of force and the influence of force are discussed.

INTRODUCTION

The destruction of polymers by break-up of mechanically stressed bonds is accompanied by formation of free radicals regardless of whether it occurs under the influence of force or thermal fluctuation [1-8]. These free radicals exhibit a short lifetime due to their activity. One of the possible ways through which these free radicals are destroyed is the oxidation reaction with the formation of chemically stable $>C=o$ groups in the molecular structure of elastomers. The existing investigations of this problem concern primarily, the thermooxidative degradation [9-20]. The mechanooxidative degradation and its role in polymer durability and the oxidative degradation in the absence of force have only recently become an object of investigation.

Therefore, we decided to investigate the mechanooxidative degradation of unfilled sulphur vulcanizates: polyisoprene (SKI-3), polybutadiene (SKD-1) and polydutadienetetryl copolymer (SKN-40), used for the investigation. They were studied both with and without inhibitors.

For SKI-3, SKD the composition of vulcanizing mixtures calculated for 100 weight parts of the rubber is: sulphur - 1.0; altax - 0.6; DFG-3 - 3.0; ZnO - 50; stearic acid - 1.0. The vulcanization mode is: T=406 K, t=30 min, P=3.5. mPa. For SKN-40 the composition of vulcanizing mixtures calculated for 100 weight parts of the rubber is: sulphur - 1.5; 2-mercaptobenzthiazol - 0.8; ZnO - 5.0; stearic acid - 1.5. The vulcanization mode is: T=416 K, t=50-60 min, P=3.5 mPa.

The products of molecular destruction were studied, mainly using the method of vibrational spectroscopy. In some cases small-angle x-ray diffraction, mass-spectroscopy, and others were used.

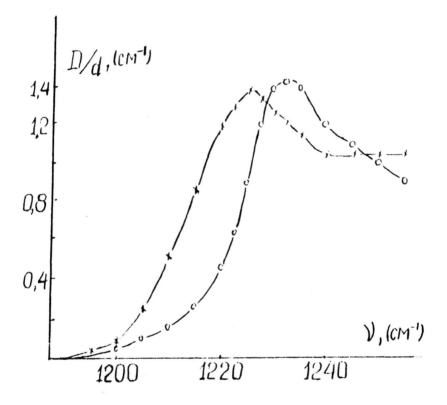

Figure 1. The influence of mechanical stress on IR spectra of polyisoprene: in the absence of load, $\sigma=0$ (1); and under the load, $\sigma=450$ MPa (2).

Consider first the elementary processes of elastomer destruction under the influence of the mechanical stress. In mechanically loaded elastomers bonds strained to maximum were observed. This was evidenced by the shift of the absorption band maximum associated with the skeleton oscillation of the elastomer molecular chain (Figure 1). According to experiments, the magnitude of the shift exhibits a non-linear dependence on the stress according to:

$$\Delta v = \alpha\sigma + \beta\sqrt{\sigma}.$$

where α and β are mechanospectroscopic factors with respective values equal to $3{\cdot}10^{-2}$ cm^{-1}/mPa and $1.5\cdot10^{-1}$ cm^{-1}/mPa.

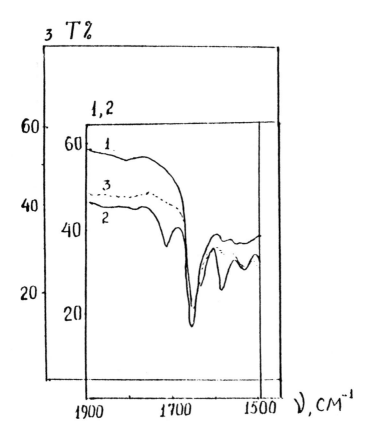

Figure 2. IR spectra of purified polyisoprene in the region 1900-1500 cm-1: under the load, σ=20 MPa, t=240 hr (2); in the absence of load (1), and of the mechanotreated sample rinsed in CCl$_4$ (3).

These bonds are given a specific significance from the standpoint of the physics of polymers since they break-up to form free radicals. The latter react with the atmospheric oxygen to transform oxygen-containing groups. These groups were treated earlier as the end groups [21-22]. The entire physics of destruction microprocesses was based on this concept [21-22].

Our experiments showed that the mechanooxidative degradation of polymers is a more complex process compared to what had been assumed earlier [22]. This is confirmed by the experiments of the mechanooxidation of polymers shown in Figure 2. It can be seen that in mechanooxidated samples a medium-intensity absorption band in the region 1720 cm^{-1} related to >C=O groups of the ketone type appears [23]. On rinsing the sample with CCl$_4$ this band completely disappears (Figure 2). This phenomenon was interpreted as follows:

1. The oxidation of mechanooxidated radicals is accompanied by breakage of molecules into fragments which were called microfragments.

2. It was also established that >C=O groups of microfragments are not at the ends of disrupted macromolecules.

Thus, our experimental results showed that the elementary act of elastomer destruction under the influence of the mechanical stress doesn't terminate in the thermofluctuational break-up of stressed bonds. This process with the oxygen participation is a more complex one resulting in the break-up of macromolecules into microfragments.

These concepts state the basis for a new approach to the mechanism of elastomer destruction under the influence of the mechanical stress.

The experimentally revealed disappearance of 1720 cm^{-1} absorption bands in oxidated samples of polyisoprene is of a universal character.

The concept of elastomer destruction under the influence of the mechanical stress as occurring according to the microfragmentary type needs a thorough analysis of the structure of both elastomers themselves and that of cleaved particles. As long as >C=O groups do not form a chemical composition of macromolecules, they must form a chemical composition of microfragments (according to the concept of microfragmentary degradation). The IR spectra of microfragments showed that >C=O groups had really formed their chemical composition.

Microfragments were dimensions using the data of gas-liquid chromatogram; these turned to be commensurate with the length of the carbon chain C$_{12}$-C$_{15}$, or 12-15 Å.

2. THE ANALYSIS OF MOLECULAR PRODUCTS OF DEGRADATION OF ELASTOMERS IN THE PRE-BREAKAGE STATE

According to the kinetic concept of stability, the essence of the process of polymer destruction is the successive break-up of stressed chemical bonds with the formation of active radicals followed by their subsequent transformation into stable >C=O groups. The measurement of the concentration of these groups with time provided the establishment of the kinetics of disruption of interatomic bonds which played an important role in the development of the basis for the process of polymer destruction. The study of the accumulation kinetics for these groups showed that the curve in coordinates had a

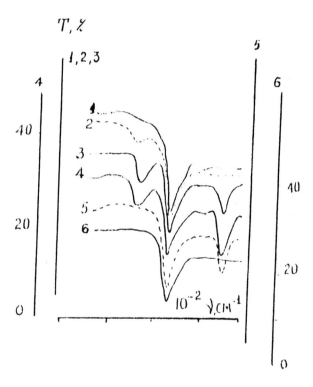

Figure 3. IR spectra of mechanically loaded polyisoprene, σ=30 MPa. The duration of exposure to stress is 75 hr (2); 230 hr (3); 670 hr (4). (1) is the unloaded sample. The same sample was rinsed in CCl₄ within 50 min (5) and 85 hr (6).

damping shape with a quasistationary portion. Figure 3 shows IR-spectra of polyisoprene samples under the load σ=30 mPa. It can be seen that a new 1720 cm⁻¹ band related to >C=O groups of the ketone type appears in a 75-hr. period. The accumulation kinetics of these groups has a quasisaturated character, too. Extremely approximated estimates show that the durability period after the concentration of the >C=O group reaches the saturation takes only 50% of the total lifetime for the samples under the load. Typical oxidative reactions with the products of the >C=O group are not detected in this prebreakage state of samples. Instead of this reaction a new channel for the oxidation reaction is generated with the formation of new products having absorption bands at 1540 cm⁻¹. The appearance of this band was also observed in oxidated sulphur-filled samples in which the vulcanization was performed at 316 K. In this case C-C bonds stand for transversal bridges.

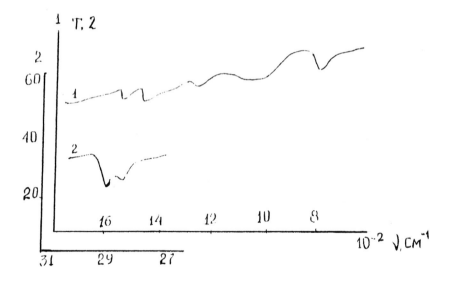

Figure 4. IR spectra of microfragments with diketone groups.

The IR spectra of these groups in oxidated samples are shown in Figure 4. The study of the accumulation kinetics for diketone groups showed that the number of these groups is increasing until the break-up of samples into two groups. It was also established that both diketone and ketone groups form a chemical composition of microfragments which is evidenced by "washing out" of the absorption band at 1540 cm^{-1}. It should be noted that the time necessary to remove microfragments with diketone groups is 80-85 hr, whereas that for ketone groups is 40-45 min.

From the comparison of both times it may be supposed that microfragments with diketone groups are significantly larger compared to those with >C=O groups. Thus, the removal of diketone groups from the elastomer matrix needs much more time compared to that for removal of microfragments with >C=O groups.

Thus, it follows that the oxidative degradation is a complex, multichannel process and the initial stage of oxidation significantly differs from the prebreakage period in degradation products.

3. IR SPECTRA OF MICROFRAGMENTS WITH DIKETONE GROUPS

To take IR spectra of diketone groups microfragments with diketone groups had to be removed from the matrix of oxidated samples. Prior to the release of diketone microfragments ketone microfragments were completely removed from oxidated samples first. Then thin films, applied to the surface of the KBr window were prepared. The IR spectra of this thin film are shown in Figure 4. It can be seen that molecular spectra of these particles, comprising diketone groups (-CO-CH$_2$-CO-) represent the superposition of spectra of both polymers themselves and their oxidated analogous. As to IR spectra in the region 3000-2600 cm^{-1}, having an absorption band of the complicated shape with a sharp maximum at v=2915 cm^{-1} and 2848 cm^{-1}, these bands relate to symmetrical and asymmetrical valent oscillations of (CH$_2$) - groups [23]. Moreover, the IR spectra of diketones have a low-intensity absorption band at 1740 cm^{-1}. This band was absent in the sample prior to washing out of microfragments. According to [23], this band is associated with oscillations of >C=O groups of the aldehyde type. The appearance of these oscillations is caused by the partial oxidation of polyisoprene during the washing out of samples in organic solvents. The intense 1540 cm^{-1} band which was lost after washing out of oxidated samples appears again in IR spectra of microfragments removed from the matrix of elastomers. Apart from this band there exist other bands at 1540 cm^{-1}, 1260 cm^{-1}, 1090 cm^{-1} and 800 cm^{-1}

revealed in the polyisoprene itself. The exceptions are the 1260 cm^{-1} and 800 cm^{-1} bands which are significantly shifted towards a long-wave region compared to those of the polyisoprene itself.

It can be seen from the analysis of microfragments with diketone groups that these particles have a very complicated structure. Apart from diketone groups there are other groups forming a chemical composition of polyisoprene macromolecules.

Thus, it follows that the elementary act of elastomer destruction in the pre-breakage state differs from that in the initial stage and terminates in the break-up of macromolecules into larger microfragment with new chemical groups (diketone).

4. On the Mechanism of Microfragmentary Destruction at the Mechanooxidation of Elastomers

Consider the mechanism of microfragmentary formation exemplified by the reaction of oxidation of purified polyisoprene under the influence of the mechanical field.

The macromolecule breaks up into 2 parts under the influence of this field to form end free radicals. On interaction with the neighboring double bond these radicals shift this bond towards the end of macromolecules (2).

$$(C_5H_3)_n - CH_2 - \overset{\displaystyle CH_3}{C} = CH - CH_2 - CH_2 - \overset{\displaystyle CH_3}{C} = CH - CH_2 - (C_5H_8)_n \tag{1}$$

$$-(C_5H_8)_n - CH_2 - \overset{\displaystyle CH_3}{C} - CH = CH_2 \tag{2}$$

On moving along the chain according to the relay-race mechanism and passing through 2-3 monomer chains these radicals react with the oxygen to form peroxide radicals $(RO_2^{\cdot})$. In case of the purified polyisoprene more close packed macromolecules are observed compared to unpurified samples.

In these systems the exchange of free radicals during the reaction occurs via neighboring molecules. The addition of the peroxide radical with a binary bond to the neighboring molecule results, in particular, in the formation of the peroxide bridge:

$$(C_5H_8)_n - CH_2 - \overset{\overset{\displaystyle CH_3}{|}}{\underset{\underset{\displaystyle O-O}{|}}{C}} - CH = CH - (C_5H_8)_2 \tag{3}$$

$$(C_5H_8)_n - H_2C - C - CH - CH_2 -$$
$$CH_3 \quad \downarrow (C_5H_8) - CH_2 + CH_3COCH = CH_2 - (C_5H_8)_2 \tag{4}$$

with the free radical released in its neighborhood.

The oxidation of the latter is accompanied by the release of the thermal energy [22]. This results in the break-up of the peroxide bridge with the formation of alcoxyl radicals. These radicals, in their turn, tear the material chain into small pieces called microfragments (4).

Thus, at close enough arrangement of macromolecules in the amorphous portion of elastomers the oxidative degradation of macromolecules develops not only along the chain but has a presumable direction via neighboring macromolecules (i.e. the exchange of free radicals occurs in the transversal direction). The development of the oxidation reaction is schematically shown in Figure 5). The similar oxidation reaction was observed in unpurified samples but the break-up of macromolecules in this case occurred intramolecularly with the formation of >C=O groups of the aldehyde type.

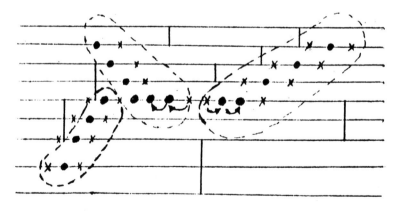

Figure 5. The diagram of the development of the degradation front of non-inhibited samples at elastomer degradation.

5. The Physical Model of Submicrovoid Formation at Microfragmentary Degradation of Elastomers

A model of submicrovoid formation differing from existing models in many aspects was created within the framework of the concept of

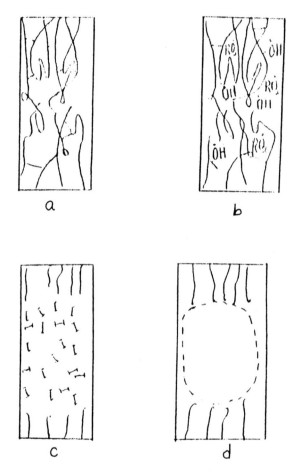

Figure 6. The diagram of submicrovoid formation at mechanooxidative degradation of elastomers: the structure of the amorphous region of elastomers (a); the initiation of chain free-radical oxidation reaction with the formation of and other active radicals (b); termination of the chain reaction of the oxidative degradation followed by the formation of microfragments (c); submicrovoids are formed, in site of the amorphous interlayer following the removal of microfragments (d).

microfragmentary degradation of elastomer macromolecules at mechanooxidation [22].

Apart from the atmospheric oxygen, the amorphous region in which all reactive groups are concentrated is known to be a microreactor in which oxidation reactions take place. The microheterogeneity of the structure of these regions is expressed clearly enough in the Weinstein or Chekh model [24]. The diagram of the oxidation reaction in the microreactor is shown in Figure 6. According to this diagram chemically active molecules providing the oxidation reaction in a chain mode are formed within each act of the reaction (diagrams a, Figure 6), resulting in the break-up of elastomer macromolecules into microfragments (Figure 6, c). Rinsing of these samples results in the removal of microfragments from the elastomer matrix and microvoids are formed instead (Figure 6,d). The presence of these microvoids was confirmed by the independent method of low-angle x-ray diffraction.

6. DIMENSIONING OF SUBMICROVOIDS IN OXIDATED ELASTOMERS BY THE X-RAY DIFFRACTION METHOD

The mechanism of generation of submicroscopic voids (submicrocracks) under the influence of the mechanical stress and their relation to the polymer structure is reviewed in good detail in [9, 22, 25-27]. The basic method used for submicrovoid detection is a low-angle x-ray scattering. There is enough evidence of the reliability of the method for dimensioning of embrionic centers of destruction (for submicrovoids) in the polymer bulk [29].

Because of vacancy (submicrovoids) formation in the bulk in the form of embrionic destruction sources on washing out of microfragments these submicrovoids can be detected using an independent method, or in other words, the dimensioning of these sources seems to be very important due to their significance.

Structural imperfections caused by these microvoids were investigated using a low-angle x-ray diffraction method (Figure 7). It follows from this figure that rinsing of oxidated samples results in the formation of submicrovoids on the x-ray diffraction scattering pattern which is evidenced by the increase of the intensity of beam scattering at $\varphi=30°$ (curve 2, Figure 7).

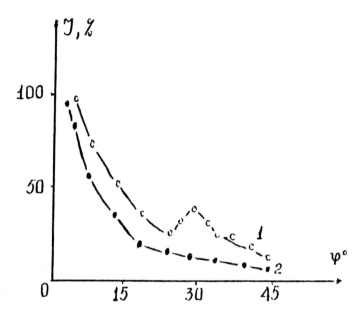

Figure 7. X-ray scattering prior to (1) and after (2) the washing out of microfragments from oxidated polyisoprene (2).

The dimensions of submicrovoids were investigated using known relations [28]:

$$D_\varphi = -\frac{\lambda}{\varphi},$$

where λ is the wavelength of the x-irradiation ($\lambda=2\text{Å}$); φ is the angle at which the maximum of x-ray scattering is observed.

The estimations showed D_φ to be equal to 150Å, i.e. the coincidence with the longitudinal dimension of the amorphous interlayer of the polyisoprene takes place. This provides a supposition of burning out of the polymer amorphous region at the oxidative degradation.

7. MECHANOOXIDATION AND DURABILITY OF ELASTOMERS

It is known that three spaced apart in time processes can be conventionally distinguished in mechanically stressed elastomers: 1) the thermofluctuational break-up of stressed macromolecules with the formation of end radicals: $RH \rightarrow R_1 + R_2$ 2) the depolymerization of microradicals with the release of volatile products [22]; and finally, 3) the oxidative degradation.

The analysis of these processes shows that the rate of the macromolecular break-up under the influence of the mechanical stress being primary acts of degradation is described by the known equation:

$$V = V_o \cdot e^{\alpha \sigma}.$$

According to our mass-spectroscopy experiments, the depolymerization of macroradicals at the mechanodegradation of elastomers is absent. As to the third process, this wasn't considered in durability estimations. Hence, the contribution of the oxidative degradation to the total picture of elastomer destruction under the influence of the mechanical stress was estimated in the present paper. The study of the accumulation kinetics of >C=O groups under the influence of the mechanical load showed that the rate of variation of the optical density dD/dt of the 1720 cm^{-1} band (of >C=O groups) vs concentration is linearly described [—]. The variation of k with the mechanical stress was determined from the slope of the diagram, and according to estimations, this stress is expressed by the exponential dependence of the type:

$$k = k_o \cdot e^{-\alpha \sigma}, or\ k = k_o \cdot e^{-\frac{E}{kt}},$$

and thus the durability takes the form:

$$\tau = \tau_o \cdot e^{E/kt}.$$

Thus, the analysis of experimental data shows that the package of constant of the degradation rate for mechanically loaded elastomers is characterized by two values: 1) the constant of the rate of the thermofluctuational break-up of C-C chemically loaded bonds; 2) the constant of the degradation rate for macromolecules into microfragments (oxidation products), i.e.:

$$k = k\sigma_2 + k_{02},$$

and we have, correspondingly:

$$\frac{1}{\tau} = \frac{1}{\tau_\sigma} + \frac{1}{\tau_{0_2}}.$$

Hence, the durability of elastomers, considering the oxidative degradation, takes the form:

$$\tau = \tau_0 \cdot \frac{e^{U_0/kt}}{1+e^{\alpha\sigma/kt}} \tag{1}$$

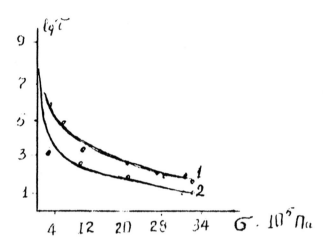

Figure 8. Time dependence of the polyisoprene stability at T=293 K: the experimental isotherm 91), and the theoretical isotherm (2).

The estimation durability curve being defined by the equation and its comparison with the experimental durability [29] are shown in Figure 8. This comparison is only quantitative in this case because in durability estimation using equation (I) experimental parameters entering this equation were determined from the linear portion of the durability curve hence, the determined uncertainty is permissible in the estimations, too. Nevertheless, the consideration of the factor of the oxidative degradation for macromolecules completely describes the previously observed experimental fact of the nonlinear dependence of elastomer durability in semilogarithmic coordinates.

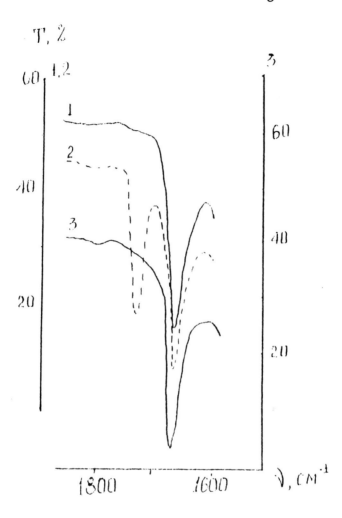

Figure 9. IR spectra of polyisoprene in the region 1800-1600 cm⁻¹: of the initial sample (1), of the sample storaged within 16-800 hr in room conditions (2), and of the sample rinsed in tetrachlorethylene (3).

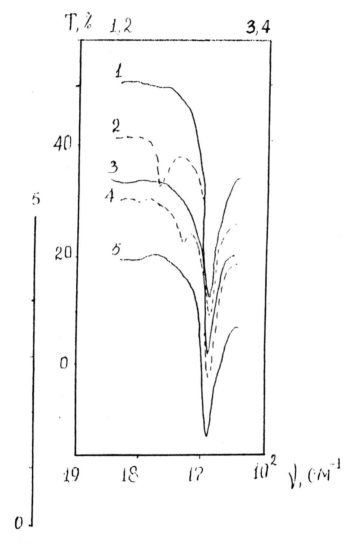

Figure 10. IR spectra of inhibited samples of polyisoprene in the region 1900-1600 cm⁻¹: of the initial sample (1) of the sample storaged within 5600 hr in room conditions (2), of the sample rinsed in CCl₄ (3), of the repeatedly oxidated rinsed sample exposed to room conditions within 5600 hr (4), of the repeatedly oxidated sample rinsed in CCl₄ (5).

Hence, consecutive experiments were devoted to the study of both the mechanical stress and the oxygen in the formation of microfragments.

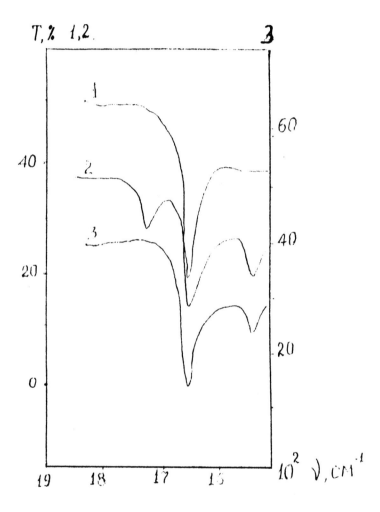

Figure 11. IR spectra of polyisoprene in the region 1800-1600 cm, of the initial sample (1), of the sample storaged within 2712 hr in room conditions (2), and of the sample rinsed in CCl (3).

First we removed the mechanical load from the source of external impacts, i.e. the experiments were performed under storage conditions in the absence of force impacts at constant temperatures and the partial pressure of the oxygen.

It should be noted that the mechanism of chain initiation in this case radically differs from that under the influence of the mechanical stress. In this case a direct attack between the oxygen and the allyl hydrogen with the formation of the medium radical $(RH + O_2 \rightarrow R^{\cdot} + O_2H)$ is observed [30].

The experiments showed that the elastomer oxidation under storage conditions at room temperature takes place according to the mechanism of macromolecular break-up into microfragments, too which is evidenced by "washing-out" of the 1720 cm^{-1} band in oxidated samples (Figure 9).

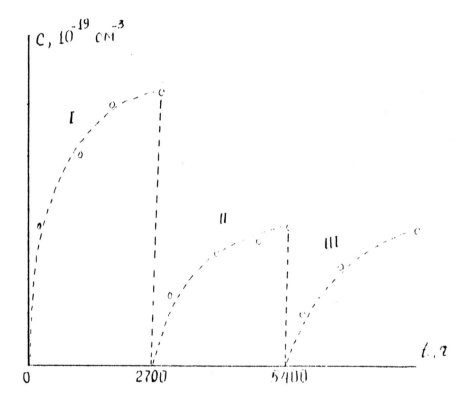

Figure 12. Accumulation of I, II and III generation microfragments with time.

The experiments on the oxidation of samples which have once passed through the oxidation-rinsing cycle turned to be of great significance. The results are shown in Figure 10. The exposition of unpurified samples within 5760 hr. results in the appearance of a new 1740 cm^{-1} band (which relates to >C=O groups of the aldehyde type). This 1740 cm^{-1} band disappeared from the IR spectrum of polyisoprene on washing out of microfragments, however the repeated oxidation of the sample, other experiments conditions being equal, resulted in the appearance of the absorption band again. However, this band differs from that of the initial oxidation stage in two aspects:

1. A new 1720 cm^{-1} band related to >C=O groups of the ketone type appears instead of the >C=O group of the aldehyde type (1740 cm^{-1}).
2. The intensity of the band in the second oxidation cycle is almost twice as low as that of the first stage.

As to the decrease of the intensity of the 1720 cm^{-1} band in the second oxidation cycle, this is characteristic of purified samples as well. The sample exposed under room conditions within 2712 hr. oxidizes and a medium-intensity 1725 cm^{-1} band appears (Figure 11, curve 2).

The repetition of the same experiment, other conditions being equal, revealed the appearance of the band with the same frequency. It may be noticed that the intensity of the 1725 cm^{-1} band is twice as low as that of the same band in the first version of the experiment. On this basis concentrations of microfragments in the first and in the second oxidation cycles were determined. The results are shown in Figure 12. It can be seen from the kinetic accumulations curve of microfragments that the sharp increase of microfragmentary concentration is observed, mainly, within the period not exceeding 2700 hr., then the process is slowly accelerating until reaching the saturation ($C = 3{,}26 \cdot 10^{+19}$ cm^{-3}). Rinsing of this sample resulted in a complete removal of >C=O groups together with their particles. This is schematically shown in Figure 12 where the concentration of >C=O groups is equal to zero.

The repeated oxidation of the sample (which passed through the oxidation - washing out cycle) has again revealed the increase of microfragmentary concentration with >C=O groups which is again decreased on rinsing of the sample and reaches zero (Figure 12, curve II).

The kinetics of microfragmentary accumulation in the third oxidation cycle is similar to that of the second cycle. However, rinsing of the sample in the third oxidation cycle results in the appearance of visually observed through holes (mesh-like) making it impossible to use it for further investigations.

Thus, summarizing the obtained results, it can be concluded that the oxidation of elastomers both during storage and under the influence of the mechanical load has a free-radical character with the following break-up of macromolecules into microfragments. Hence, in this case and in the case of mechanooxidation $>C=O$ groups form a chemical composition of microfragments, and do not represent end groups in disrupted macromolecules. Remarkable in this respect are experiments performed with different oxidation - rinsing cycles which are peculiar in that concentrations of the second and the third generations of microfragments are similar to each other being much lower compared to that for microfragments of the first generation. How can such a significant difference in concentration for microfragments of the first and subsequent generations be accounted for? Unfortunately, direct data accounting for this experimental fact are absent, nevertheless this can be explained in terms of the existence of weak bonds in polymers for the latter are responsible for the initiation of destructive processes, and many aspects of the low-energy stability of polymers are associated with them [29]. The growth of $>C=O$ groups in the first version of the experiment seems to be explained by the existence of weak bonds, and their break-up results in the extensive accumulation of oxygen-bearing groups in the polymer matrix. The number of these groups is limited, so they are completely consumed during the first version of the experiment. In subsequent versions equivalent bonds stay within polymers, so their oxidation results in the appearance of similar microfragments.

8. DEGRADATION OF ELASTOMERS IN PHYSICALLY CORROSIVE LIQUIDS

The destruction of elastomers as a physico-chemical process is associated with force surmounting of interatomic or interchain bonds under the influence of force factors. The most important here is the detection of changes in the chemical structure of polymers as the consequence of the post effect at the macromolecular break-up. It is known that the chemical nature of polymers isn't changed on destruction along interchain bonds [31]. This can be exemplified by the polymer break-up in physically corrosive liquids with the formation of molecular solutions.

As to cross-linked polymers, they swell in these media, and their stabilities turn to be much lower compared to those of unswelled samples. It is confirmed in [32,33] that physically corrosive liquids cause reversible changes in polymers. Hence, the break-up of chemical bonds isn't observed in this process. At the same time in real conditions polymer materials are in a constant contact with physically corrosive media hence, irreversible processes are observed. This is especially dangerous in physically corrosive liquids in which the polymer is strongly swelling. In these cases the significance is attached to the type of interchain bonds. C-C cross bridges, for instance, are more stable compared to sulphuric -C-S-C bonds. The swelling of cross-linked polymers in organic solvents depends on solubility parameters of both

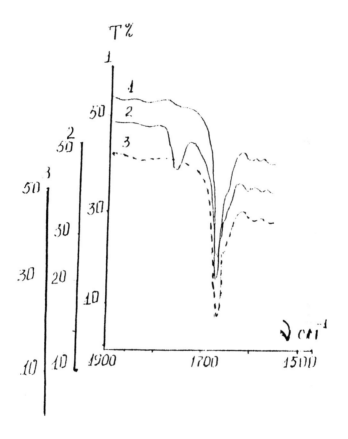

Figure 13. IR spectra of polyisoprene in the region 1900-1600 cm^{-1}: of the initial sample (1), of the sample oxidated in CCl$_4$ within t=5 min (2), after washing out of the oxidated sample (3).

polymers and the solvent [31]. In case of coincidence of these parameters a significant swelling of the polymer is observed. The mechanism of penetration of physically corrosive liquids into the polymer is described in more detail in [31-33], in which it is noticed that the transfer of low-molecular liquids in the polymer matrix occurs, mainly, via both the activated diffusion and submicrocapillar fluxes.

Not going into details of swelling processes of polymers, note that the above process is accompanied by the formation of internal stresses [34]. The role of these stresses in oxidation-destructive processes is rather poorly described [31-33]. In connection with this we've studied the degradation of the cross-linked polyisoprene in organic solvents. CCL$_4$, benzene, etc., in which the polyisoprene is significantly swelled were used as organic solvents. Experimental results are shown in Figure 13. It can be seen from this figure that samples of polyisoprene are strongly oxidized and form >C=O groups. This is manifested in the appearance of the spectrum of the 1740 cm^{-1} absorption band which is however washed out on rinsing of the sample (Figure 13, curve 2). This washing out of the band provided by oxygen-bearing groups is the evidence of their close association with macromolecules of elastomers, and the formation of chemical composition of microfragments disrupted from macromolecules during the oxidation reaction. Thus, it follows that the oxidation of polyisoprene is independent of the way the oxidation is initiated. In all the above cases (under the influence of the mechanical stress, during the storage or in organic solvents) the break-up of macromolecules into microfragments is observed.

Methodologically, there is a circumstance which complicates the fixation of microfragments in the elastomer matrix during its oxidation in organic solvents. As a matter of fact, the oxidation of polyisoprene in organic solvents is accompanied by the break-up of its macromolecules into microfragments which are simultaneously released from the elastomer matrix during the oxidation. This results in that >C=O groups can not be always fixed in the polyisoprene matrix following its stay in solvents. Hence, a special procedure for the fixation of >C=O groups in elastomers following their oxidation in physically corrosive liquids has been developed. The essence of this procedure is as follows. Specimens were placed into small crucibles with a predetermined quantity of the solvent. The volume of these solvents was selected so, that specimens could simultaneously react both with the liquid and the air (i.e. rinsing conditions for oxidated polymers were deliberately violated). Using this oxidation mode provided the fixation of >C=O groups at elastomer oxidation in corrosive liquids. Similar to mechanooxidation, these groups also form a chemical composition of microfragments. In this case the reaction is initiated by medium sulphuric radicals which are formed under the influence of internal stresses. The comparison between the experiment on mechanooxidation and that on the oxidation of elastomers in the absence of force factors (storage conditions) shows that the model of

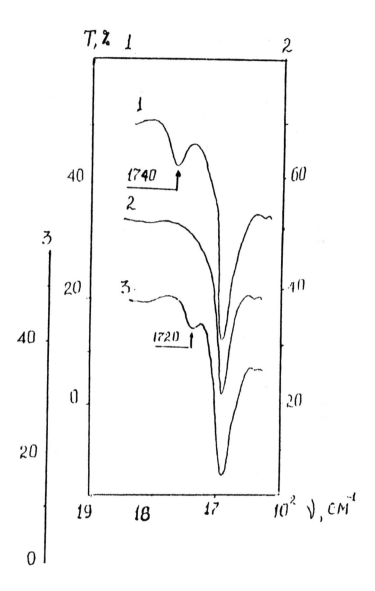

Figure 14. IR spectra of polyisoprene: oxidated in CCl$_4$ (1), following the washing out of the oxidated sample (2), of the repeatedly storaged sample in room conditions within 5600 hr (3).

physical destruction is based on the microfragmentary break-up of macromolecules which is independent of the way the reaction is initiated.

Another important circumstances resulting from these experiments is that aldehydes (having 1740 cm^{-1} absorption bands in the IR region) stand for products of elastomer oxidation of the >C=O group in organic solvents.

It can be seen from the diagram of microfragmentary formation that the oxidation in pure samples occurs with the participation of interchain exchanges between free radicals resulting in the formation of >C=O groups of the ketone type. The oxidation of elastomers in organic solvents may serve as a working model for the influence of interchain distances monitoring the reaction. It has been already mentioned that these elastomers are strongly swelling in these media corresponding to the increase of interchain distances, and each molecule acts as an isolated system. The oxidation reaction in these cases flows purely intramolecularly. It is known from the reaction flow diagram that using this oxidation channel results in that aldehydes stand for oxidation products. This is favored by experiments on the repeated oxidation of samples which passed through the stage of oxidation in organic solvents. The results are shown in Figure 14. It is seen that polyisoprene is subjected to oxidation in CCl$_4$ resulting in the formation of >C=O groups of the aldehyde type. The washing out of this band results in a complete transparency of the sample in this region. The repeated oxidation of this sample during the storage and the influence of the mechanical stress result in the formation of >C=O groups of the ketone type. The convincingness of the hypothesis of the formation of >C=O groups is beyond any doubt, since it is confirmed by the experiment performed using one and the same sample placed in different media, and having different interchain distances, correspondingly.

Thus, the experiments on elastomer oxidation performed both during the storage and in physically corrosive liquids show that the mechanical load doesn't play any significant role in the process of the macromolecular break-up into microfragments, except for the initiation of the reaction.

9. ELASTOMER DEGRADATION STUDY BY THE MASS-SPECTROMETRY METHOD

The study of polymer mechanodegradation under high vacuum conditions shows that this process is associated with the release of monomer chains [21-22]. This is an additional evidence of that the break-up of chemical bonds occurs under the influence of the mechanical stress. In a rough approximation, these monomer chains could be called microfragments. It may be supposed, of course, that microfragments forming under high vacuum conditions differ from those forming in the air.

Thus, the analysis of [24] shows that microfragments can be formed at elastomer mechanodegradation (with the oxygen participation). To unambiguously determine the mechanism of formation of these particles, the experiments were performed under high vacuum conditions. The mass spectroscopy analysis of elastomer break-up products is usually used for this purpose.

It should be noted that the study of elastomer degradation by the mass spectroscopy method is associated with many difficulties caused by the high elastic strain of elastomers themselves. Hence, their study by a direct loading in the vacuum chamber of mass spectrometer performed similar to that described in [35-37] is impossible. That's why the mechanodegradation of polymers was performed using a ball mill, its schematic diagram being similar to that of the vibromill used in [38, 39]. It should be noted that the experiments on mass spectra investigation of volatile products at mechanodegradation of elastomers is the most convenient and the direct method of microfragmentary detection (if any) under high vacuum conditions.

The experiment on elastomer mechanodegradation was performed using a vibromill which was directly attached to a supply system of the mass spectrometer MSH-4. These experiments showed that the mechanical breakup of polyisoprene is not accompanied by the release of volatile products at room temperature, whereas a break-up of macromolecules into microfragments was observed during the experiments on elastomer mechanodegradation performed using the same sample [40-42].

Thus, the comparison of experimental results on elastomer mechanodegradation performed under vacuum with those in the air shows that it's the oxygen, not other external forces that are responsible for microfragmentary formation, since otherwise volatile products should be formed under vacuum as well as influenced by the mechanical stress similar to the case of PMMA [29].

Thus, summarizing experimental results, following remarks can be made:

1. The oxidation of elastomers is accompanied by the break-up of macromolecules into small fragments called microfragments, irregardless of the way the reaction was initiated (by force or using any other external factor). The microstructural analysis showed the length and the chemical composition of microfragments to be independent of the way the reaction was initiated, and is commensurate with the length of the carbonic chain - $C_{12} \div C_{15}$.

2. New oxidation products are observed in a deep stage of oxidation namely, β - diketones. Diketone groups were found to form a chemical composition of microfragments as well. The size and the molecular structure of diketone microfragments significantly differ from those with >C=O groups.

3. The durability isotherm (lg $\tau=\sigma$) is shown to be described by two processes, namely the thermofluctuational process, and the oxidative process, the latter becoming decisive with the decrease of the stress.

4. Different products of oxidation are formed in the purified and unpurified elastomers: ketones are formed in the purified elastomer, whereas aldehydes are formed in that inhibited by other impurities. Aldehydes are formed in swelled elastomers in the absence of impurities as well. This is interpreted in terms of both the competition within and the interchain transfer of free valency. In pure elastomers the interchain transfer becomes dominant, and the ketone formation is observed, whereas inhibited elastomers (with impurities) are characterized by the formation of aldehydes.

5. A physical model of submicrovoid formation regarded as embrionic sources of destruction at mechanooxidation and oxidation of elastomers in the absence of the influence of force is generated.

REFERENCES

1. S.N. Zhurkov, S.Ye. Bresler, E.N. Kazbekov and Ye.E. Tomashevskii: *Zhurnal teoret. fiziki*, v. 29, vyp. 2, 358-365 (1954).
2. S.N. Zhurkov, V.A. Zakrevskii and E.Ye. Tomashevskii: FTT, v. 6, vyp. 6, 1962-1964 (1962).
3. P.Yu. Butyagin: *Vysokomol. Soed.* A, v. 9, N 11, 136-138 (1967).
4. V.A. Meshnevskii: *Vysokomol. Soed.* B, v. 11, N 1, 44-49 (1969).
5. T.N. Kleinert and J.R. Norton: *Nature*, v. 6, 334-336 (1962).
6. K. Ulbert and p.Yu. Bytyagin: Dokl. AN USSR, v. 149, N 5, 1194-1196 (1963).
7. G.V. Abyagin and P.Yu. Butyagin: *Vysokomol. Soed.*, v. 7, N 8, 1410-1414 (1965).
8. P.Yu. Butyagin, V.F. Drozdovskii, D.R. Razgon and I.V. Kolbanov: FTT, v. 7, vyp. 3, 941-943 (1965).
9. G. Kaush, *Destruction of polymers*, transl. from English (ed. by S.B. Ratnev), Moscow, Mir, 1981, 440.
10. J.L. Morand: *Rubber Chem. and Technol.*, v. 47, N 5, 1094-1115 (1974).
11. Yu.B. Shikov and Ye.T. Denisov: *Vysokomol. Soed.* A, v. 19, N 6, 1244-1249 (1977).
12. Ya.M. Slobodin, V.S. Maiorova and N.A. Smirnova: *Vysokomol. Soed.* v. 6, N 3, 541-543 (1964).

13. V.V. Pchelintzev and Ye.T. Denisov: *Vysokomol. Soed.* A, v. 27, N 6, 1123-1136 (1985).
14. V.V. Pshelintzev and Ye.T. Denisov: *Vysokomol. Soed.* v. 25, N 5, 1025-1041 (1983).
15. N.L. Panteleeva, Ye.A. Ilyina and S.M. Kavun: *Vysokomol. Soed.*, v. 26, N 7, 1471-1476 (1984).
16. Ye.A. Ilyina, S.M. Kavun and Z.I. Tarasova: *Vysokomol. Soed.* B, v. 17, N 5, 388-390 (1975).
17. N.M. Lamaev, V.A. Kurbatov and A.G. Maksimovich: *Vysokomol. Soed.* A, v. 23, N 2, 375-380 (1984).
18. A.A. Popov, P.Ya. Rapoport and G.Ye. Zaikov, *Oxidation of oriented stressed polymers*, Khimia, Moscow, 1987, 229.
19. A.S. Kuzminskii, N.M. Lezhnev and Yu.S. Zuev, *Oxidation of rubbers and vulcanizates*, Moscow, Gosnauchtekhizdat, 1957, 379.
20. K.B. Piotrovskii and Z.N. Tarasova, *Ageing and stabilization of synthetic rubbers and vulcanizates*, Moscow, Khimia, 1980, 264.
21. I. Dekhant, R. Dantz, B. Kimmer and R. Shmolk, *IR spectroscopy of polymers*, trans from German (ed. by E.F. Oleynik), Moscow, Khimia, 1976, 471.
22. V.R. Regel, A.I. Sluzker and E.Ye. Tomashevskii, *Kinetic nature of stability of solids*, Moscow, Nauka, 1974, 560.
23. L. Bellami, *IR spectra of complex molecules*, Moscow, Foreign languages publishing house, 1974, 560.
24. N.M. Emmanuel and A.L. Buchacenko, *Chemical physics of ageing and stabilization of polymers*, Moscow, Nauka, 1982, 267.
25. P. Griffith: Trans. Roy. Soc. 1921, 221A, 163.
26. A.F. Ioffe, *Physics of crystals*, Leningrad, Gosizdat, 1929.
27. S.N. Zhurkov, V.A. Zakharevskii, V.Ye. Korsukov and V.V. Kuksenko, *Mechanism of submicrocavity generation in polymers under the load*: FTT, v. 12, N 10, 2857-2864 (1970).
28. R. James, *Optical principles of x-ray diffraction*, trans. from English (ed. by M.P. Shaskolskaya), Moscow, Foreign languages publishing house, 1950, 572.
29. G.M. Bartenev and Yu.S. Zuev, *Stability and destruction of highly elastic matters*, Moscow-Leningrad, Khimia, 1964, 373.
30. K.B. Piotrovskii and Z.N. Tarasova, *Ageing and stabilization of synthetic rubbers and vulcanizates*, Moscow, Khimia, 1980.
31. Yu.S. Zuev, *Destruction of polymers under the influence of corrosive media*, Moscow, Khimia, 1972, 228.
32. P.A. Rebinder and A.P. Pisarenko: Dokl. AN USSR, v. 73, N 1, 129-132 (1950).
33. A.A. Tager, *Physics and chemistry of polymers*, Moscow, Goskhimizdat, 1963, 582.
34. A.A. Popov, S.D. Razumovskii, V.N. Parfenov and G.Ye. Zaikov: Izv. AN USSR, ser. Khimia, N 2, 282-287 (1975).

35. V.R. Regel, T.M. Muinov and O.F. Pozdnyakov: *FTT*, v. 4, N 9, 468-473 (1962).
36. V.R. Regel and T.M. Muinov: *Vysokomol. Soed.*, v. 8, N 5, 841-845 9196).
37. Kh. Khabibulloev and D. Saidov: *Rubber and vulcanizate, vyp.* 3, 41-47 (1982).
38. V.A. Radtsig, V.S. Pudov and P.Yu. Butyagin: *Vysokomol. Soed.*, v. 9, N 6, 414-415 (1967).
39. V.A. Malchevskii, O.F. Pozdnyakov, V.R. Regel and M.S. Falkovskii: *Vysokomol. Soed.*, v. 8, N 9, 2078-2085 (1971).
40. D. Saidov, *Structure and physical properties of elastomers*, coll., Moscow, TzNIITENneftekhim, 1981, 79-81.
41. D. Saidov, N. Narzulloev and K.B. Nelson: *Vysokomol. Soed.*, v. 29, N 5, 1028-1031 (1987).
42. D. Saidov: *Dokl. AN Tadj. SSR*, v. 29, N 2, 88-91 (1986).

FIRE AND HEAT SHIELD MATERIALS BASED ON SULFOCHLORINATED POLYETHYLENE

M.A. Shashkina*, R.M. Aseeva,
A.A. Donskoy* and G.E. Zaikov
*Institute of Aviation Materials, Moscow
Institute of Chemical Physics, Russian Academy of Sciences
117334, Kosygin Str. 4, Moscow, Russia

INTRODUCTION

The increase of aircraft safety depends on an objective, detailed study of accident causes. Therefore an analysis of flight parameters and crew working during start, flight and accident was performed.

Flight information recording systems (FIRS), which record the main parameters of a flight on tape, are objective witnesses of accident. The FIRS are usually placed in metal container with fire shield cover to block and decrease heat flow from the surroundings. A container with a fire shield cover should protect the recorded information from thermal degradation, mechanical destruction and humidity. This container or "black box" would provide the information regarding the cause of any accident that may occur.

Recently, similar information systems have been used in railway, water and underground transport systems.

This paper considers the physico-chemical properties of efficient fire and heat shielding materials (FHSM) for use in FIRS. The data on the development of FHSM based on sulfochlorinated polyethylene are shown. Great attention has been paid to the analysis of mechanism of fire protective action for the developed materials.

2. Physico-chemical Properties of Efficient FHSM for Use in Flight Information Recording Systems

There are several types of FIRS, differing by construction and amount of the recorded information. The limit of acceptable temperature inside the container (for example, 60°C for polyester tape and 350°C - for metal one) changes depending on the type of material used in making the tape. First construction of FIRS comprised an external fire and heat shielding cover and a metal blow resistant container, into which a heat isolation layer was placed. Glass reinforced plastics possess a heat protective properties but their blow resistance is not high. If an aircraft should crash, such covers would break. Therefore it is necessary to create elastic, blow resistant covers, providing a safe heat shield for the device and after its falling from height. Moreover, the task of the decrease of device overall size and its weight put forward the question of the development of highly efficient FHSM, allowing to protect the information carrier without using internal thermal insulation layer.

The background for making the heat shield materials, including the elastic ones, was developed by rocket-cosmic technology research. However, the FHSMs for FIRS differ from heat shielding used in air-space vehicles in working conditions. The temperature is lower (1100°-1600°C) while the heat exposure time is much longer (up to 30 min.).

FHSM's performance consists of two stages, an active and a passive one. In case of FIRS the active stage is stipulated by the influence of flame on the surface of cover for an extended period of time. After the end of the exposure to external heat, the passive stage starts. Here, the heat energy stored in the cover during active stage becomes the source of heat. For efficient work of FHSM it is important to dissipate a maximum amount of energy input for the transformation of material during the first stage. During the second stage, however, the objective of the cover material is to direct heat emission into the surroundings.

In the FIRS construction, used in the former USSR during the 60s, the external cover was made of rubber-fabric material with the binder based on silicon polymers. An other multilayer system used for this purpose consisted of an intumescing cellulosic material, impregnated with products of amine condensation with aliphatic aldehydes. A thermoplastic with aluminum foil was used as external protective layer. The following materials were recommended for internal thermal insulation: mineral and glass wool, vermiculite or asbestos paper [1].

Efficient and good for practice system has been suggested for the protection of fuel tans [2]. The scheme of layer disposition in the construction is shown at Figure 1. FRG patent [3] describes the way for making fire and heat shielding container, in which polymeric for is used as internal thermal insulation. The containers for the storage of magnet carriers and other susceptible to overheating materials are shown in the works [4,5]. Here, there are used both an external layer for heat shield and an internal thermal insulation. USA patent [6] proposes FHSM and the way for its making. The material consist of a few layers. An external layer is ablative cover. This layer swells up and decomposes at the heating, absorbing a large amount of heat energy. Thus, the influence of fire on lower layers of material is decreased.

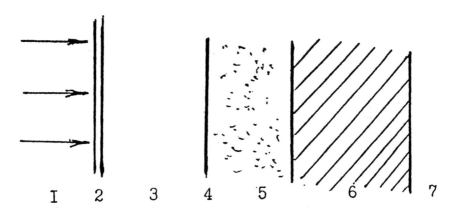

Figure 1. Scheme of layer disposition for fire and heat shield of fuel tanks: 1-flame; 2-thin steel cover; 3-vacuum or air; 4-reflecting surface; 5-thermal insulation; 6-tank wall; 7-fuel.

Technology for FHSM making included the following procedure: a thin layer of material with porous or cellular structure was formed by pressing. Then the layer of thermoplastic with low decomposition temperature (93°-204°C) ant the intumescent coating were prepared.

The above described FHSM's represent complex systems with various mechanism for heat dissipation. A thorough understanding global chemical and physical processes controlling mass and heat transfer in the systems is essential to develop new efficient FHSM's.

The heat transported from flame or other external source of thermal energy to the surface of polymeric material can increase its temperature to decomposition one. The resulting fuel vapor mixes and reacts with the oxidizer from the surroundings to initiate ignition and to provide flame spread. Therefore, the problem of efficient FHSM development is narrowly connected with the one of flame retardancy of polymers.

Heat transfer is usually carried out from hot bodies to cold ones by thermal conduction or convection. Thermal conduction takes place in both gas and solid phase while convection is only through the gas phase. At large fire the main mode for heat transfer from flame or incandescent body to FHSM surface is radiation. The external radiation flux transferred to any body surface may be divided in three components: reflected, absorbed and permeated ones [7-9]. Radiation heat flux, passed through the cover, is brought to the surface of the protected object. Absorbed part of radiation energy is spent for the transformation of the cover material [10, 11]. The amount of the energy absorbed by polymer material depends on the level and spectral characteristics of the radiation flux, absorption and reflection abilities of the material. Reflection ability of the cover depends on such factors as of the state of the surface, beam incidence and wave length of flux input, refraction indices of the mediums.

Radiation energy is dispersed when wave length is much smaller than the size of irregularities on rough surface. The chemical composition of the material effect transmissivity and absorption coefficient of the cover. If absorption coefficient is small, radiative heat flux will penetrate the material thickness and its heating will be retarded. However, if absorption coefficient is large, radiative energy absorbed on the surface of the material will rapidly increase its temperature to critical value of the decomposition beginning. In this case the maximum temperature of the material is observed at its surface.

For efficient FHSM development it is important to increase reflecting power of the material. However, polymers and other organic compounds have low reflectivity [10]. Reflecting power can be increased by a special modification of material surface or by the introduction in polymer of fillers which are capable to reflect radiative energy. Oxides of different metals can be used as such fillers.

The depth of heated layer of the material and time for its heating up to critical temperature depend on chemical nature of polymeric material, its thermophysical properties and power of incident heat flux [11]. The surface layer and heating time of this layer decrease with heat flux power or with the decrease of thermal inertia (lrc). Cellular polymers with low value of thermal inertia have a higher thermal insulating properties. However, there are the danger for their faster ignition and flame spread over the material surface in comparison with monolythic counterparts.

Polymeric FHSM's are composite materials. They include many components for a different special purposes (fillers, plasticizers, flame retardants, etc.). The transformation processes of the components and the material on the whole at the action of heat energy are complicated. The effects of these processes on mass and heat transfer at the combustion of heat shielding cover are not well understood. However, it is quite evident that the endothermic processes, accompanied by heat absorption, and the processes of the energy dissipation into the surroundings are preferable.

Physical endothermic processes are phase transitions of the substances: the melting of crystal ones, the evaporation or the sublimination, the transition from solid state to viscous liquid one.

Chemical reactions with the disassociation of molecular bonds are also accompanied by heat absorption. However, total thermal effect of the substance decomposition depends on energetics of following elementary reactions.

The thermal decomposition of polymers in dependence on their chemical nature can be carried out by different paths.

Two large groups of polymers are usually distinguished: the polymers, which degradate at high temperatures practically completely with the formation of low molecular volatiles, and the polymers, which form a carbonized product or nonvolatile residue.

In spite of the differences in the decomposition mechanism (for example, reverse depolymerization or random scission of the main macromolecular chain with the formation of smaller fragments), decomposition of polymers of the first group is characterized by endothermic effect. Polymers of the second group show the tendency to exothermic reactions of the decomposition. Such reactions, as elimination of side substituents in main chains with the formation of conjugated double bonds, cyclization, intermolecular cross-linking, recombination are exothermic ones. They can cause following carbonization of nonvolatile residue. The bond scission and volatilization of low molecular products also take place at the decomposition of the polymers. Therefore, total heat effect for low temperature step of the decomposition is often endothermal or close to neutral one due to the compensation by the heat of exothermic reactions. More high temperature steps of the decomposition for char-forming polymers are carried out with heat release, as well as the reactions with oxygen participation.

Carbonizable thermoplastics at the first steps of the process can expand and swell up, promoting the formation of foamed char. Morphological structure of char layer affect polymer flammability and heat shielding properties of the material. On one hand, surface char layer due to high absorptive ability accumulates a large amount of heat energy. At the growth of temperature on the surface of char layer the heat quantity emitted in the surroundings as well as transferred to lower material layers increases [7]. On the other hand, foamed char has a lower density and thermal conductivity than initial polymeric material. As the result, thermal insulating ability of foamed char layer begin to play dominant role in the protective of action of FHSM.

The char strength has the importance for effective heat shield. The change of cover volume at constant sizes of protected surface, thermal contractions and volatilization of the decomposition products are the cause for the appearance of strains in FHSM and partial destruction of the cover.

Chars with large-cell structure are usually friable and ineffective as thermal insulation.

Understanding physical and chemical processes, participating in the formation of firm foamed char on cover surface, is necessary to develop the new efficient FHSM's.

Theoretical models for intumescent char formation are discussed in papers [12,13]. Mathematical models for pyrolysis of carbonizing polymers at unidimensional heating are represented by researchers [14,15]. Paper [15] considers one-stage and double-stage char formation at pyrolysis of polymers. The influence of surface heterogeneous chemical reaction and the formation of porous structure of char due to the appearance and the motion of bubbles of volatile decomposition products was analyzed.

The suggested models are based on some simplifying assumptions. In particular, complex kinetics of pyrolysis is described by only one global temperature dependence with Arrhenius law. It is supposed that substance transformation into char begins after the reaching some threshold temperature, which is specific for the polymer.

Heterogeneous reaction of char oxidation leads to the generation and the accumulation of heat inside the cover. Therefore, polymers decomposed to nonsmouldering residue and unflammable gases are preferable for FHSM development.

Flame retardancy of polymer materials can be carried out by physical and chemical methods [16].

Physical ways for the action on the combustion process, being interesting for FHSM development, were mentioned earlier. These ways must help to decrease heat input, to increase the dissipation of heat into the surroundings and to make worse mass transfer of fuel to the zone of the combustion reaction.

The directed change of polymer structure, the composition and component ratio of material may be considered as chemical measures for the regulation of the combustion of polymer material. For the development of flame retardant polymeric materials the next ways are important: the synthesis of noncombustible or hardly combustible polymers; the surface and voluminal chemical modification of materials; the application of efficient flame retardants; the aimed blending of polymers; applying fire shield coatings; the combination of the above mentioned ways. As result the resistance of material to ignition should increase. The rate of flame spread over the polymer surface and heat release one should decrease.

Besides these important indices, fire safety of polymeric materials is characterized by toxicity of combustion products and the level of smoke formation.

Stable flame combustion of polymers includes the totality of mutually adjusted gas and solid-phase processes. This adjustment is stipulated by level and rate of heat flux incident from flame to polymer surface. It can be expressed by concept of Spoulding mass transfer number, B, taking into account radiative component and heat losses or combustion efficiency [17].

Mass transfer number, B, is connected directly with rate of gasification, $\dot{m}''$, by the following expression:

$$\dot{m}'' = h \ln(1 + B)/c_p,$$

where h - convective coefficient of heat transfer; c_p - heat capacity of the polymeric material.

Mass transfer number, B, can be represented as follows:

$$B = [\{Y_{ox} \Delta G_c (1 - X_R) X_A / r\} - c_p (T_s - T_o)] / \{L(1 - E)\}$$

Here Y_{ox} is mass content of oxygen in the surroundings; ΔH_c is the heat of full combustion; X_A - combustion degree; X_R - radiation energy part; T_s, T_o - temperature of the surface and t surroundings; L - the gasification heat of polymeric material; E - the combustion efficiency.

E, taking into account radiative heat losses from the material surface, can be expressed as follows:

$$E = (\dot{q}_e'' + \dot{q}_{fr}'' - \dot{q}_{rr}'') / \dot{m}'' L,$$

where $\dot{q}_e''$ is external heat flux; $\dot{q}_{fr}''$ - reverse radiative heat flux from flame; $\dot{q}_{rr}''$ - radiative heat losses from the material surface. Other modes of heat losses at material combustion can be taken in the consideration.

Polymers with low values of combustion heat and high one of gasification heat are relatively lower combustible materials. Therefore, inorganic polymers wit silicon, nitrogen, fluorine and other non-carbon elements in the main chains of macromolecules are a special interest.

Thermal stable polymers with aromatic or heterocyclic structure are also important [16]. However, such polymers are yet expensive. Moreover, their processing into articles is often difficult Hydrolytic stability of the polymers is not enough in some cases.

Another above mentioned ways for flame retardancy of polymeric materials are more accessible.

Thermal stability of polymer can effect its flammability characteristics. For example, it was found that anomalous structural units in main chains of PMMA decreases thermal stability of the polymer. The gasification rate increases by 25% and flame spread one over the surface of material increases by 4 times, comparing with more stable sample [17].

The definite scaling relationships between thermal parameters and the fire response of idealized polymeric materials were found by R.E. Lyon [18a]. However, the similar interconnection between thermal stability and flammability indices for flame retardant polymer is not of universal character. Besides the gasification rate, the content of combustible degradation products is very important. Flame retardant polymeric materials are often characterized by more lower decomposition temperature than initial polymers.

The application of flame retardants with aimed functions (fillers, plasticizers, foaming agents, etc.) is the most wide-spread approach to fire protective materials development.

Flame retardants represent the substances, containing such chemical elements as B, P, halogens, nitrogen, different metals. Mechanism of flame retardant action depends on the nature of an element, chemical composition of the substance and polymer, the conditions of heat influence on the system. A number of the studies is devoted to this question [18,19].

Flame retardants are distinguished in the dependence on the zone of leading influence as the substances with the mechanism of the action in gas phase and the substances, which acts in condensed one. Halogen-containing compounds are corresponded to typical flame retardants of gas phase action. Volatile products of their decomposition play the role of inert dilutors of flame medium or inhibitors of gas phase chain reactions.

Bromine-containing flame retardants are more efficient than chlorine-containing ones. However, temperature affects the efficiency of halogen-containing flame retardants. At large fire, when the temperature exceeds 12000-1300 C, bromine shows oxidative properties and catalyzes chain gas phase reactions. Thus bromine can perform unwanted influence on the combustion process on the whole [20].

Mechanism action for most of phosphorus-containing flame retardants is mainly the condensed phase one. Phosphorus-containing substances can decompose at the combustion of polymeric materials to oxides or acid derivatives of phosphorus. Dense surface layer of polyacid compounds of phosphorus creates physical barrier for oxygen and for the diffusion of the combustible products of pyrolysis into the zone of flame reaction. Phosphorus-containing compounds promote the increase of carbonized residue. Besides, they change the direction of oxidative reactions for char residue, decreasing their exothermicity and suppressing the material smouldering. Nitrogen-containing substances increase the effect of phosphorus flame retardants.

Borates, silicates, different salts of alkali metals can also form glass protective layers [21].

It should be noted that the flame retardancy of polymeric materials is often carried out by mixed, condensed and gas phase mechanism.

A great number of the substances was suggested as flame retardants and smoke supressants [22]. Besides simple substances with high content of flame retardant element (red phosphorus, as example), some mineral natural and inorganic synthetic compounds were suggested: asbestos, talcum [23], bismuth and molybdenum, selves or their mixtures [24], oxide and hydroxide of aluminum [25], ammonium salts [26], carbides of metals [27].

Fillers for flame retardant polymeric material are often separated as inert and active ones. However, this separation is relative. it is necessary to take into account concrete temperature conditions in the material use. Asbestos fiber at temperature below 900°C is usual inert dilutor with low thermal conductivity. At temperature o 900°-1400°C asbestos produces crystallizational water. Above 1400°C endothermic process of reforming chemical structure of asbestos is carried out. The chemical interaction between asbestos components and polymer decomposition products is possible.

Flame retardant effect of inorganic fillers is usually considered as result of the decrease of fuel part of the material or the change in thermophysical properties of one. However, it could be seen a surprising coincidence: some inorganic flame retardants are catalysts for the reactions of polymerization and polycondensation in polymer synthesis. Therefore, it should be supposed that they can accelerate the reactions of cross-linking and char-forming at polymer combustion. TiO_2, Co_2O_3, Al_2O_3, aluminum phosphates are catalysts for the reactions of dehydrocyclization and aromatization of hydrocarbons [29]. The action mechanism for high effective flame retardants of intumescent type is connected with catalysis of char forming reactions in polymeric system [29].

The number of known flame retardants is great. However, a definite demands limit probable choice for the concrete material development. Main general demands for flame retardants of polymers are following: the compatibility with the polymer; the stability at temperature of material processing; non toxicity and corrosion inactivity; the absence of the negative effects on technological properties; mechanical and other characteristics of the material.

The following factors were taken into account to select the polymer for FHSM development: the accessibility of raw basis; developed industrial production; relatively low price; inherent decreased combustibility. Elastomers are preferable for impact-resistant FHSM development.

Halogen-containing polymers are very interesting for the purpose. Native industry produces a different chlorine-containing polymers: homopolymers and copolymers of vinylchloride or vinyliden chloride; chlorinated butylrubber, polyisoprene and polyvinylchloride; chlorinated and sulfochlorinated polyethylene, etc. Additional halogenation of polymers enlarges the application spheres of the products [30,31].

Polymer chlorination can be carried out in medium of organic solvents, in water suspension and in solid phase by gas or liquid chlorine action. Sulfonation can be performed simultaneously with chlorination. The process is initiated by visible, UV, ionizing radiation or radical initiators (organic peroxides, hydroperoxides, diazo compounds and metalorganic ones) [3]. The properties of chlorinated polymers depend on the nature of initial polymer, chlorination degree, the way and the conditions of making [30].

Chlorinated and sulfochlorinated polyolefines are widely used for the different purposes.

Let us consider the peculiarities of the making and the properties of the products.

3. THE PROPERTIES OF CHLORINATED AND SULFOCHLORINATED POLYETHYLENES

Halogenation of polyolefines shows some features, stipulated by chemical passivity of initial polymers and also the presence of amorphous and crystalline parts in their structure. The reactions of halogenation can be accompanied by the degradation and cross-linking of the main chains of macromolecules [32, 33]. The process is mainly carried out in solvent medium or in water suspension. At room temperature, polyethylene is not soluble in common solvents. In the dependence on the crystallinity degree, PE begins to solve in chlorinated aliphatic and aromatic hydrocarbons at temperatures above 53°C. PEHD is solved fully in CCl_4 at the temperature close to boiling one. PELH solubility is observed at higher temperatures (80-100°C) [30]. The solubility of chlorinated PE depends on chlorine content in the polymer. The solubility of chlorinated PE increases with chlorine content up to 30% mass, then it decreases. The polymer becomes insoluble at 50-60% mass of chlorine content [30].

USA patent describes PE chlorination in CCl_4 solution with radical initiators [34]. Using chlorobenzene with higher solving ability allows to increase the reaction rate [35]. The different radical initiators [36,37] and their mixtures [38], as example, dinitril of isobutyric acid with benzoyl or laurylperoxides, chloral dihydroxyperoxide together with acetylcyclohexylsulfonyl peroxide, were suggested. ICI Company (Great Britain) developed technology for PEHD chlorination in CCl_4 suspension or acetic acid one by using metal chlorides [30]. Liquid chlorine was suggested as chlorinating gent [39]. However, gaseous reagent is mostly applied. The product with chlorine content of 40-44% mass. was obtained by suspensional way using gaseous chlorine [40]. The different surface-active substances were used for water suspensional chlorination of PELD [41]. Chlorinated PE's with chlorine content of 25-50% represent amorphous or low-crystal polymers having the tensile strength from 8 to 17 MPa and relative elongation at tensile of 350-800%. At larger range of chlorination (from 14 to 70%) the various products are obtained: from thermoplastis to friable solids [42]. The modification changes the structure of final polymer. Polymer crystallinity breaks at chlorination in solution stronger than at suspension one. Polymer crystallinity is partially remained in the last case at chlorine content up to 55% mass [32]. It was found that $-CCl_2-$ groups are absent in the macromolecule chains of chlorinated PE. There are not also -CHCl-CHCl- vycinal groups when chlorine content is lower 55% [32]. At first, amorphous areas of polymer take part in chlorination reaction, then more regulated ones are implicated. The reaction of hydrogen substitution by chlorine proceeds statistically resulting random distribution of chlorine in polymer chain.

Submolecular structure of polymer changes from spiral for PE to spherolytic (monocrystal) at chlorine content of 8% mass. Increasing chlorine content up to 14-18% accompanied by the appearance of short fibrils. At chlorine content of 22-30% mass. packs and globules are observed [32]. Completely amorphizated structure of polymer is observed at chlorine content above 55%. Physical state of PE changes with chlorination degree in the following way [41]:

polymer state	chlorine content, % mass.
semielastical thermoplastics	10
thermoplastical elastomer	20
thermoplastical, viscous elastomer	30
thermoplastical elastomer	40
semielastical thermoplastic	50
rigid thermoplastic	60
friable polymer	75

Glass transition temperature for the polymers with chlorine content of 25-35% lies in the limits from -20° to +30°C. The value increases up to 100-180°C when the chlorine content is of 38-73% mass. Softening temperature change is of parabolic character with the minimum for 35-40% mass. chlorine content [30]. Chlorinated PE is resistant to acids, weak alkali and salt solutions. It shows inherent decreased combustibility. At flame action the product does not melt and is charring. The char residue yield increases with chlorine content [31].

Numerous works [41-45] are devote to the study of chlorinated PE thermal decomposition. The process mechanism is more complicated comparing with initial PE one. The decomposition of chlorinated PE is multistage process, including both the rupture of main chain bonds and the substituent elimination to form polyene fragments for following transformation into cross-linking and aromatic structures.

Caoutchouc-like products of PE sulfochlorination are very interesting for FHSM development. The making of sulfochlorinated polyethylene (SCPE) is carried out by barboting mixture of gaseous chlorine with sulfurous acid anhydride (in molar ratio of 2:1, preferable) through PE solution in CCl_4 at 70-75%C. Peroxides or asobisisobutyronitrile are used as radical initiators.

Native industry produces 6 trade marks of SCPE based on PEHD [46]. Foreign assortment has more variety of the products. Du Pont Company, USA, produce 8 trade marks of SCPE based on both PEHD and PELD.

SCPE, produced by native industry, with content of 26-36 %CL and of 0.85-2%S begin to decompose at 115°-150°C. The tensile strength is about 15.0 MPa and relative elongation at tensile is 300-350%[42]. The mechanical properties are changed with increasing sulfochlorination degree to chlorine content of 59% and sulfur one of 4.4% [30,49]. The effect of structural factors on the properties of PE modified was analyzed in the study [47]. It was found that the branchings in polymer do not affect the properties of final product. Molecular weight and the molecular mass distribution of the initial polymer are more important factors. The effect of degree of chlorination on relaxation properties and phase state for SCPE was discussed in reference [49]. SCPE with chlorine content up to 59% and sulfur one up to 4.4% was obtained on the basis of PEHD (ρ=966 kg/m^3). Spectrum of relaxational and phase transitions stipulated by different modes of molecular motion was determined by methods of dynamic mechanic spectrometry, linear dilatometry and DSC. The polymer crystallinity decreases with Cl content. Full amorphisation of SCPE is observed for chlorination degree of 30± 2% mass. Shearing modulus decreases with decreasing polymer crystallinity (Figure 2, curve 1). Shearing modulus was measured at temperature correspond to low-temperature boarder for high elastic state of amorphous part in the polymer. Bending curves for the dependence of the density and glass transition temperature (Figure 2, curves 2 and 3) on chlorination degree correspond to full amorphization of polymer. DSC data prove the conclusion. Endopeak on SCPE thermograms, connected with crystal phase melting, changes its form, shifts to the lower temperature and decreases with increasing chlorine content up to 30% mass.

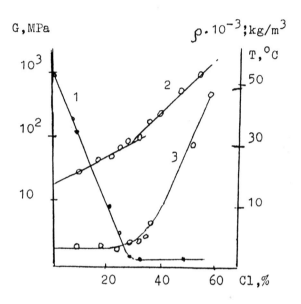

Figure 2. The influence of chlorine content in SCPE on the change of shear modulus (1), density (2) and glass transition temperature (3) of the polymer.

Sulfochlorinated PE possesses important advantage in comparison with chlorinated product: better ability to vulcanization. Optimum sulfur content in polymer to obtain impact-resistant product is about 1.5%. SCPE are characterized by high light-, ozone-, oil- and petrol resistance and also inherent decreased combustibility [50]. Non-vulcanized SCPE's are stored for a year without changing their properties [51]. SCPE vulcanizates attract the attention by the high resistance to various atmospheric influences and ionizing radiations [30]. In tropic conditions, they retain their properties for a long time, enduring the influence of mold and microorganisms.

High static strength of SCPE vulcanizates is observed even in absence of the reinforcing fillers. SCPE fragility temperature varies in the range from -55° to 60°C. SCPE vulcanizates show the satisfactory tear resistance and attrition one, retaining at increased temperature. They are characterized by high resistance to frequent tensiling and bending loads, to the decrepitation and heat aging. Maximum working temperatures for the most of the articles based on SCPE fall within the range of 130-160°C. At 100°C on the vulcanization conditions and vulcanizing agent type [52,53].

Vulcanization of elastomers is complex process of transformation of linear macromolecules into net structures with relatively rare cross links. As result, the product loses the solubility and the thermoplasticity properties, but acquires a high elasticity, strength and a number of new valuable qualities.

The different reactionable atoms and groups can participate in the net structure formation: chlorosulfonic groups, chlorine and hydrogen atoms, unsaturated double bonds formed by polymer dehydrochlorination [54,55]. Chlorine atoms in β-position to chlorosulfonic groups are mostly labile and able to catalytic dehydrochlorination.

The presence of different reactionable atom and groups in SCPE allows to apply a different organic and inorganic compounds as vulcanizing systems. The next substances were suggested for this purpose: metal oxides, sulfur and sulfur-containing compounds, peroxides, disocyanates, nitrogen-containing bifunctional compounds, polyatomic alcohols, etc. [54-58]. Vulcanizating systems, including oxides of lead and magnesium, are the most effective for SCPE [54]. At present, combined vulcanizating systems, which include oxide or salt of polyvalent metal (10-50 parts mass.), organic acid (2-10 parts), sulfur-containing compounds as vulcanization accelerators (0.5-10 parts), are widely used [30]. It has been found that the sulfur-containing accelerators of vulcanization in the systems are true cross-linking agents.

It is supposed that metal oxides make no direct influence on chemical transformation of SCPE macromolecules. Metal oxides participate in net structure formation as sorption surface, dispergator of true cross-linking agent and absorbant of gaseous products [30]. According to another conception, metal oxides promote the transformation of chlorosulfonic groups into more polar ones of basic salt type [58]. During polymer heating chlorosulfonic groups in presence of moisture and activators of acidic type (for example, fatty acids) or vulcanization accelerators (for example, thiurams) hydrolize with forming HCl and $-SO_2H$ groups. The latter participate in the formation of metal-sulfonate bonds. Accelerator of sulfonic vulcanization react with chlorosulfonic groups. The cross-linkage or the voluminous side substituents are appeared. The side substituents are able to associate with each other and with polar groups on the surface of metal oxide, also. The links of polymer with metal oxide formed as the result of adsorption or hemosorption of macromolecules, are stable at usual temperatures of material exploitation.

Thiazoles, thiurams, dithiodicarbamates are used as sulfur-containing accelerators of vulcanization for SCPE. Aminoaldehyde accelerators call scorching of SCPE during its processing [54,55]. Guanidines are used only as secondary accelerators in cooperation with thiazoles [30]. High vulcanizing activity for SCPE is shown by salts of hexamethylenediamine and diatomic acids (hexamethylene diammonium sebacinate, for example, [57]) or the products of their condensation [59]. In this case magnesium oxide increases the cross-linking degree, and sulfur additives accelerate the process of culcanizate in absence of the additives and increases the strength of vulcanizate at presence of magnesium oxide. The combination of magnesium sulfide with diphenylguanidine and 2-mercaptoimidasoline is very effective [61,62]. Polyatomic alcohols (for example, pentaerythritol [63]), chlorinated aromatic substances (as example, p-xylene hexachloride [64]) are used as vulcanizing systems. The regimes for SCPE vulcanization by different vulcanizing agents are discussed in ref. [65]. The composition formulations based on SCPE for making materials with high resistance to heat or light-ozone aging, aggressive media or water action were described [66].

Technology of SCPE processing does not differ from that for other elastomers used for a long time in rubber industry. Thus, the equipment, equal for the industry, is suitable. It should be noted that nonvulcanized SCPE is more thermoplastic than natural or many other synthetic caoutchoucs. Therefore, it needs no preliminary plasticization [30,48]. Making rubber mixtures on roller or in the mixers is accompanied by sufficient heat release which can lead to subvulcanization of the compositions. To avoid this process it is necessary to make mixing as quick as possible. Rubber mixtures based on SCPE are formed satisfactory by pressing, extrusion, calendering or pressure molding. Vulcanization of SCPE mixtures is carried out at temperature of 120-150°C. A higher temperatures of processing can lead to the appearance of porosity, shells and other defects on the surface of the material. Vulcanization of SCPE compositions can be performed in press or by sharp vapor at pressure up to 1.8 MN.m^{-2}. In the case there is wide plato of vulcanization and SCPE mixtures are resistant to scorching.

4. FIRE AND HEAT SHIELD MATERIALS BASED ON SULFOCHLORINATED POLYETHYLENE

Chlorinated or sulfochlorinated PE itself or in the mixture with other caoutchoucs and thermoplastics were used for the development of flame retardant materials. For example, there were described flame retardant, impact resistant compositions based on chlorinated and sulfochlorinated PE with chlorine content of 25-45% mass. [67]. The compositions and the compounds for making fire shield coatings, including additionally halogen-containing flame retardants, synergists and inorganic fillers, have been suggested [68-73]. SCPE vulcanizates with low chlorine content are flame retardant ones, but they resign in this property chloroprene compositions. New types of SCPE with high chlorine content are equal to chloroprene or exceed it by flame retardancy rating. The analysis of the properties of chlorinated and sulfochlorinated PE lead to the conclusion that at the development of impact resistant FHSM for FIRS, it is expedient to use an industrial SCPE with chlorine content of 27% and sulfur one of 1.3-1.5% mass.

The combined vulcanizing system included magnesium oxide, sulfur vulcanization accelerator and some other aimed components has been used to obtain the compositions with high technological properties, resistant to scorching.

The composition formulation contains silicon and antimony oxides, chloroparaffin of CP-470 trade mark with chlorine content of 47% mass. Silicon oxide (White ssot) is an active filler for SCPE. The filler allows to obtain rubbers with high mechanical properties and, in particular, with high resistance to tearing and frequent bending loads. Chloroparaffin is an efficient plasticizer and simultaneously flame retardant for non-colour SCPE rubbers. Antimony oxide is synergyst for halogen-containing flame retardants, increasing fire shielding effect. The compositions were obtained by rubber milling. The samples of vulcanizates are formed by pressing at temperature of 140°C and pressure of 5 kg.cm^{-2}.

Optimum formulation for composition has been found as the result of the variation of component content and the study of final FHSM properties. Optimum ratio polymer and vulcanizing system is 100 to 27.5 parts mass.

To understand mechanism of fire shielding action of the coating for FIRS, the effect of used metal oxides and chloroparaffin on the flammability and thermal properties was investigated in detail.

The table 1 shows the results illustrating the effect of the nature and the content of metal oxide in SCPE vulcanizate on the values of limiting oxygen index and flame spread rate over sample surface.

Limiting oxygen index was measured by standard method of GOST 12.1.044-84 and also for flame spread in horizontal or vertical upward directions. For these cases the sample sizes and the reactor ones, the rate of nitrogen-oxygen flow were corresponded to standard conditions of LOI determination. The influence of thermal insulating support on LOI and flame spread rate was estimated. LOI for nonvulcanized SCPE equals 22.5%. The table 1 shows that the orientation of the samples, the direction of flame spread and support presence affect the flammability indices. The sample thickness at testing was equal to 4-5 mm. The effect of sample thickness on LOI value for SCPE vulcanizates of optimum formulation is shown in table 2. LOI is practically constant for sample thickness of 3.5-7.3 mm. LOI decrease at thickness lower 3.2-3.5 mm is stipulated by relative decrease of heat losses from flame in the surroundings. LOI values for candle burning and for flame spread over horizontal surface of the samples without support are close. The sufficient increase of LOI and the change of flame spread rate are observed when support is used (Table 1). This fact confirms the presence of essential heat flowing through condensed phase and the support. The rate of flame spread along the sample surface increases with oxygen content in the surroundings. The most essential change of LOI is observed when candle combustion is replaced by upward one. In latter case the rate of flame spread increases with time. In the rigid conditions of the combustion the SCPE vulcanizates without fillers are able to ignite and to burn in air media (LOI is equal 20.5%).

The flammability of SCPE rubbers decreases at the introduction of metal oxides. Antimony oxide is more effective as flame retardant than silicon oxide at filler content up to 30 parts mass. However, the influence of metal nature is being leveled with the increase of metal content up to 50-70 parts, if LOI is determined by standard method of GOST 12.1.044-84. In this case the rate of flame spread at the extinguishing limit for SCPE filled vulcanizates is 2-4 times lower than that for non-filled ones. High efficiency of antimony oxide as flame retardant in comparison with silicon oxide (white soot) is observed in upwards flame spread mode (Table 1). It should be noted that the additional introduction of chloroparaffin into the compositions with high content of filler (antimony or silicon oxides) improves technological properties of ones, but it affects insignificantly LOI values of SCPE vulcanizates. However, the positive influence of chloroparaffin on LOI increase was observed at the combination of these fillers.

SCPE rubber of optimum formulation, included all the above mentioned components, shows the highest LOI values [Table 2) and the lowest rates of flame spread over the surface in comparable conditions.

Fire shielding properties of the materials for FIRS were estimated by the method of heat blow. The device with one-side heating material was used (Figure 3). Planar heater with temperature of 1100°C was contacted with the surface of sample (size of 80x80mm, 10 mm thick). Temperature of the protected surface was recorded automatically by thermocouple. The junction of thermocouple was situated in the center of back side of FHSM sample on the protected surface (Figure 3). The rate of temperature change of the protected surface and maximum temperature in the definite time moment were estimating criterions. The used methodics reproduces the simplified model of working fire shielding cover. The methodics allows to perform screening FHSM and to select fastly the most efficient one. At testing fire shield properties fast ignition of the sample is observed. Flame embraces all surface contacted with the heater. Burning out of surface layer is accompanied by charring polymeric material. Carbonization front is moved inside the sample. Reaching 450°C at shielded surface, charring material is observed through all thickness. It was found by the help of X-ray analysis that there is practically no antimony in the carbonized residue. All other metals (Mg and Si) are retained as oxides.

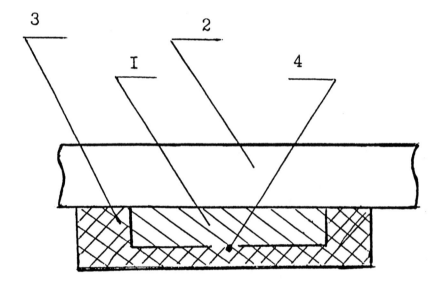

Figure 3. The device scheme for determination of fire and heat shield properties of materials: 1-sample; 2-heater; 3-thermal insulation; 4-thermocouple junction.

The testing of fire shield properties SCPE rubbers shows that silicon oxide allows to decelerate the heating rate largely than antimony oxide. However, the additional introduction of chloroparaffin into the composition with Sb_2O_3 positively affects the decrease of temperature on the protected surface (Figure 4).

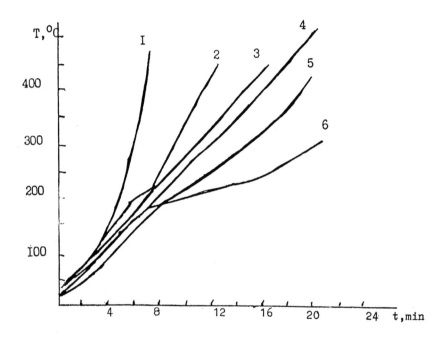

Figure 4. Fire and heat shield properties of the composition based on SCPE: 1-vulcanized SCPE; 2-SCPE with 50 mass parts of Sb_2O_3; 3-SCPE with 50 mass parts of SiO_2; 4-SCPE with 50 mass parts of Sb_2O_3 and CP-47-; 5-optimum formulation; 6-the composition with chlorobromoparaffin.

At the combination of two metal oxides and chloroparaffin the efficiency of fire shielding action of the coating increases. The substitution of chloroparaffin by chlorobromoparaffin with equal total halogen content and Br=25% mass. allows to decrease largely the heating rate of the protected surface (Figure 4, curve 6). However, this effect appears at test duration, exceeding 6-8 min. Apparently, at lower duration of the testing there is not enough time to form the layer of foamed coke required thickness. At this stage the heat shield is determined by thermophysical properties of the initial material and its thermal inertia.

The mechanism of heat shielded action of the cover is complex. To clear up the physical and chemical processes in this mechanism, the thermal properties of SCPE vulcanizates were studied. Heat flows from flame to the FHSM sample surface were also estimated [74].

The thermooxidative decomposition of initial polymer and SCPE vulcanizates represent multistage process at dynamic heating (7 degree/min). Metal oxide nature affects the temperature range and the rate of volatilization, as well as the yield of char residue. The transition temperature of initial polymer into high elastic state is observed above 100°C. Endopeak with the maximum at 125° C on the DTA curve for the sample corresponds to this transition. Small mass losses (about 4% mass.) below 200°C are stipulated by the volatalizaton of low molecular substances, which there are in initial polymer. The intensive decomposition of SCPE is observed in temperature range of 230°-400°C. The process proceeds according to the first order reaction with the rate constant of $k=4.3 \times 10^8$ exp (-94700 J/RT), min $^{-1}$.

The elimination of side substituents in the main chain and the degradation of chlorosulfonic groups with the formation of sulfur dioxide and hydrogen chloride are carried out during this stage. Vulcanized SCPE decomposes intensively at temperature over 182°C. Mass losses at heating up to 182°C do not exceed 2%. The first stage of the decomposition in the range of 182°-360°C is practically thermoneutral one unlike high temperature stages. It should be noted that despite the decrease of temperature of SCPE decomposition beginning after vulcanization of polymer, the rate of the process at the first stage is retarded. Activation energy of the decomposition of SCPE vulcanizate increases up to 100 kJ/mole. The reaction order become the second one (Table 3). It is connected with conformational and spatial changes in net polymer structure, the decrease of macromolecule mobility after vulcanization. Magnesium oxide included into vulcanizing system affect the change of macromolecular mobility. Additional introduction of fillers (silicon or antimony oxides) causes the increase of SCPE rubber stability. The opposite effect is observed at using plasticizers - chloroparaffin or other compounds (Table 3). It is known that effective activation energy of polymer decomposition depends on the initiation reaction. The data obtained point out different character of molecular interaction of polymeric matrix with surface atoms or groups of metal oxides. In our opinion this interaction causes the inactivation of labile SCPE centers, initiating the degradation of cross-linking polymer. Plasticizer promotes the polymer decomposition by the deterioration of the interaction with the surface of metal oxides. However, it should be taken into account that chloroparaffin and antimony oxide are reactionable compounds. Chloroparaffin of CP-470 mark decomposes at temperature above 130°C at dynamic heating. Its LOI is 25% [75]. The interaction of antimony oxide with hydrogen chloride (the product of SCPE and CP-470 dehydrochlorination) leads to the formation of oxychlorides and volatile antimony chloride [76,77]. The initial intermediate product - antimony oxychloride SbOCl - is instable and decomposes at 240-280°C according to the following reaction:

$$5SbOCl \rightarrow Sb_4O_5Cl_{2(s)} + SbCl_{3(g)}$$

The following decomposition of $Sb_4O_5Cl_2$ at higher temperature (400°-570°C) limits the transfer of $SbCLl_3$ into gas phase [76-78].

Similar antimony oxybromides are formed in the presence of bromine-containing compounds and Sb_2O_3. $Sb_4O_5Br_2$ was found in the condensed phase of the combustion products for polyolefine compositions with synergetic mixtures of Sb_2O_3 and bromine-containing flame retardant [79].

The effectivity of flame retardant action depends on the ratio of Sb:Hal in the composition, the interaction character Sb_2O_3 and halogen-containing compound and also polymer nature. The factors affect the amount of $SbHal_3$, coming to gas phase. $SbHal_3$ is efficient inhibitor of flame radical reactions. Characteristic feature of thermooxidative decomposition of SCPE rubbers is the proceeding of the secondary reactions of polymer charring and the oxidation of carbonized residue at temperature over 500°C. Polymer transformation into carbonized structure with fragments of condensed aromatic cycles is carried out after dehydrochlorination and the appearance of unsaturated π-conjugated double bonds in macromolecules.

Magnesium and silicon oxides do not practically affect the char residue yield. Antimony oxide increases the char residue one, if take into account the elimination of antimony from the system as volatile derivatives. It is interesting to look for the process, how atomic ratio of chlorine and antimony in the compositions affects flammability indices of SCPE rubbers. The compositions, containing only antimony oxide, have Cl:Sb ratio from 7.46 (for 30 mass. parts of Sb_2O_3) to 1.6 (for 70 mass parts of one).

The additional introduction of chloroparaffin in a high filled system increases the ratio to 2.16, coming close to the value of 2.26 for SCPE composition with 50 mass parts of Sb_2O_3. It is known that the most large effect of flame retardancy is reached at optimum ratio of Cl:Sb need the formation of $SbCl_3$. There is no surprise that the highest change of LOI is observed for the system with 30 mass parts of

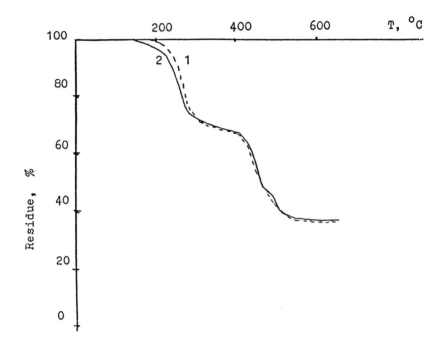

Figure 5. Thermogravimetric curves for FHSM composition containing CP-470 (1) and chlorobromoparaffin (2) at air heating with the rate of 7 degree/min.

Sb_2O_3. Additional introduction of chloroparaffin into highly filled composition does not provide optimum ratio of Cl:Sb=3 and does not lead to LOI increase. At the same time, the ratio Cl:Sb comes close to optimum one (3.04) if chloroparaffin is used in the composition with 50 mass. parts of Sb_2O_3. Thus, the efficiency of fire shielding action of FHSM increases (Figure 4, curves 2 and 4). The most large fire shield effect was observed when chloroparaffin substituted by chlorobromoparaffin in SCPE composition. However, atomic ratio Hal:Sb decreases to 2:1 in this case.

The analysis of FHSM thermogravimetric curves shows that only low temperature stages of the decomposition have an essential differences. Slow decrease of weight of FHSM sample with chlorobromoparaffin begins over 130°C. At 220°C (the start of intensive decomposition) weight losses are 6% (instead of 2% for the composition with CP-470).

The additional stage with neutral heat effect and low weight loss is observed in the range of 330°-403°C. The yield of nonvolatile residue is similar at 490° and 535°C (Figure 5).

The obtained data allows to conclude that the efficiency of FHSM depends on the reactions in condensed phase, chemical nature and the amount of flame retardant transferred in gas phase.

At halogen surplus HHal and $SbHal_3$ appears in the gas phase. At halogen deficiency $SbHal_3$ is a main volatile product. Nonreacted antimony oxide is the inert dilutor of combustible part in the condensed phase. It is known that $SbHal_3$ is more efficient flame retardant of radical gas reactions than HHal. The latter acts usually as inert dilutor in the gas phase. Antimony halides ($SbCl_3$ and $SbBr_3$) shows in the flame double function: 1. they are the source of HHal; 2. they form antimony monooxide, SbO, which participates in the catalysis of recombination reactions of active radical (H, O, OH) in the flame through intermediate particles, such as SbOH [80]. Spatial zone and the stay time of inhibitor particles in flame increase due to different boiling temperature of antimony halides (223°C for $SbCl_3$ and 288°C for $SbBr_3$).

To determine the mechanism of flame spread over FHSM surface the temperature distribution in the combustion wave has been measured. Oxygen concentration in nitrogen-oxygen media was equal to 46%. The details of the measurement and the calculation of heat flu x are given in paper [74]. The next thermal-physical properties of SCPE composition were used for the treatment of experimental results:

thermal conductivity λ=0.33 W/m.K;
specific heat capacity c_p=1.3 kJ/kg.K;
thermal diffusivity   a=1.6x10^{-7} m^2/s;
density ρ=1550 kg/m^3.

Temperature reaches 830°C in the flame edge at the distance at 3mm from the sample surface. Maximum temperature of the flame outside the edge is 1230°C temperature of the surface under the flame edge is 620°C. However, in a far distance from the flame edge, temperature on the surface increases up to 960°C because of charring and char oxidation.

The flame edge was determined by the position of maximum temperature gradient in the gas phase. Figure 6 shows the distribution of heat flows on the FHSM surface.

Heat balance at the combustion has taken into account heat flows transferred to the surface from flame by convection through the gas phase ($\dot{q}_g''$), by heat conductivity through condensed phase ($\dot{q}_{cs}''$) and also by radiation ($\dot{q}_{fr}''$). Heat losses are probable because of irradiation and reflection of heat energy by FHSM surface:

$$\dot{q}_l'' = \dot{q}_{rr}'' + \dot{q}_{ref}''$$

Thus, heat balance can be presented by the equation:

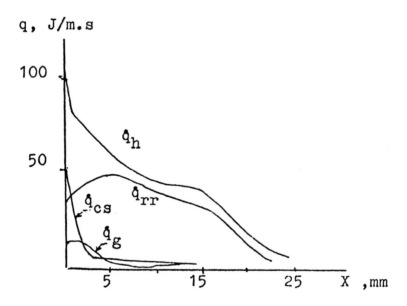

Figure 6. The distribution of heat flows at downward flame spread on the surface of SCPE vulcanizate sample containing 30 mass parts of Sb_2O_3.

$$\dot{q}_s'' = \dot{q}_g'' + \dot{q}_{cs}'' + \dot{q}_{fr}'' - \dot{q}_l''$$

Incident heat flow before at pyrolysis front is expended for the change of enthalpy of surface material layer, $\dot{q}_h''$. Radiation energy absorbed by FHSM was estimated by the difference between $\dot{q}_h''$ and $(\dot{q}_g'' + \dot{q}_{cs}'')$. As it is seen from Figure 6, the main contribution into surface layer enthalpy near the flame edge is made by radiation heat transfer from flame by conduction through condensed phase. Far from the edge this contribution is mainly made by the radiation from flame. Heat transfer through the gas phase near the flame edge by convection is 9.5% from the total heat flow. The contribution of radiation energy is 49.1% and that of heat transfer through condensed phase is 41.4%. Probably, a large contribution of radiation in total heat transfer is connected with carbon spot formation during FHSM combustion. The change character for radiative heat flow in preflame zone (5mm from flame edge) shows sufficient heat losses due to reflection of energy and irradiation from material surface. Metal oxides with high reflection coefficients for IR range [7] increase reflective ability of FHSM.

Data obtained shows the complex mechanism of heat shielding action for developed FHSM. Both physical and chemical factors are the importance. The processes proceeding in the condensed and gas phase, at the interface affect the heat and mass transfer at FHSM combustion, fire shield functions of the material.

The investigation allowed to develop the rubber-like impact-resistant FHSM of high efficiency. Coatings for FIRS, making from developed material, were tested in the conditions modelling natural ones.

During testing the following indices were estimated: the resistance to distributed static load of 1000 kg at 30 min exposition. This index was determined for two axes of device; the resistance to blow of load of 250 kg, falling from 1m height. The touch square was equal 1.6 cm². Blow overloads approach to 1000 units. During all these tests there were no breaches of FHSM completeness. The information tape was fully protected and the record can be reproduced without any disturbances.

The influence of thermal impact of 1000°C with 15 min exposition of all device surface was estimated for FIRS construction of two types: 1) usual construction with internal heat insulation and 2) the construction without internal insulating layer. The results of testing first type of FIRS construction are shown in Table 4. The developed coatings and traditionally used rubber-fiber FHSM were compared. The following parameters were controlled:

temperature at titanium FIRS box;
time of reaching maximum temperature;

After 15 min exposition of thermal load, temperature was measured both at surface of titanium box and inside FIRS container. Maximum temperature inside the container was measured also.

Table 4 shows the advantages of the developed materials based on SCPE. Testing device of the second type demonstrated the increase of temperature inside box to 150°-180°C after 15 min. heat action. However, temperature increased after finishing external heat action. Temperature achieves 190°-280°C depends on SCPE composition after 20-22 min. Thus, high level of heat shield properties of developed materials is observed in this case, also.

The developed FHSM allow to refuse from complex FIRS constructions with using internal thermal insulation.

Table 1

The Influence of Metal Oxides on Flammability of PESC Vulcanizates

Indices	Direction of flame spread	Metal oxide content, pph of PESC						
		Sb₂O₃				SiO₂		
		0	30	50	70	30	50	70
LOI,%	↓	27,8	37,8	40	42,7	34,8	39,5	42,5
$V_{fs} \cdot 10$, mm/s	↓	!2,01/ !/28	0,56 /38	0,52/ /40	0,85/ /43	!0,7/ !/36	I,I8/ /40	0,73/ /43
OI , %	↑	! 20,5	34,5	35,8	37,8	!27,%	29,5	30,5
OI , %	<--	! 26,8	38	40,5	43	!33,8	38	40
$V_{fs} \cdot 10$, mm/s	without asbest	0,66/ 1/27	0,5/ /38	0,7/ /4I	0,53/ /43	0,90/ 1/34	0,84/ /38	0,45/ /40
OI , %	<-- with asbest	! 31 I,64/ !/3I	43,5 -	45,5 I,82/ /46	48 0,68/ /48	!4I I,75/ 1/4I	48,5 2,5/ /49	52,8 I,35/ /53
$V_{fs} \cdot I0$, mm/s								

denominator—oxygen content in the atmosphere

Table 2
The Influence of Specimen Thickness on Oxygen Index of FHSM with Optimal Formulation

Thickness, mm	Oxygen index ,%			
	Direction of flame spread			
	$\downarrow$	$\uparrow$	$\leftarrow$ without asbest	$\leftarrow$ with asbest
1,3	41,8	3I,5	41,5	43,0
3,2	44,0	40,5	44,7	58,0
3,5	46,5	$>42^X$	47,0	57,5
7,3	47,0	$>42^X$	--	57,0
8,5	·46,5	$>42^X$	48,0	62,0

Table 3
Kinetic Parameters of Thermooxidative Decomposition for PESC Compositions

Sample	Temperature range, $^{\circ}C$	$E_{эф} \pm 2$, KJ/mol	z, min^{-I}	n
PESC	230–400	94,7	$4,3.10^8$	I
PESC	I82–362	II0,0	$I,6.10^{II}$	2
vulcanizate(I)				
$I+SiO_2(50)$	195–380	130,0	$I,2.10^{I3}$	2
$I+Sb_2O_3(50)$	220–400	I67,2	$8,1.10^{17}$	2
$I+Sb_2O_3+ \ +CP-470$	II0–385	I45,0	$7,2.10^{I4}$	I,5
$I+Sb_2O_3+ \ +SiO_2+CP-470$	I20–400	I40,0	$2,8.10^{I3}$	2

Table 4

Testing Results of Thermal Impact Stability for FIR with Different Fire and Heat Shield Coatings

Shield Material	Temperature, $^{\circ}$C				
	on the outside titanium FIR box				inside FIR box
	T_{max}	Time to T_{max} ,min	after 15 min testing		T_{max}
Silicon Rubber Textile	890	7^x	810^x	80^x	420
Material based on PESC	380	34	248	25	250

x- high temperature action has been stoped after achievment

of temperature inside FIR box of 80°C

REFERENCES

1. Pat. USA 3934066.
2. J.A. Mansfield, S.R. Ricitiello, L.L. Fewell, *J. Fire and Flammability* vol. 6, N4, 492 (1975).
3. Pat. Germany 1270792.
4. Pat. USA 3888557.
5. Fire vol. 69, N854, 135 (1976).
6. Pat. USA 3816226.
7. M. Derybers, *Practical Application of Infrared Beems, Energy*, Moscow (1959), in Russian.
8. Yu.V. Polezhaev, F.B. Yurevich, *Heat Shield, Energy*, Moscow, 91976), in Russian.
9. I.V. Margolin, N.P. Rumyantsev, *Principles of Infrared Tecniques*, Military Publishers, (1957), in Russian.
10. J. Lecont, *Infrared Radiation, Phys. Mathem.* Publishers, Moscow (1958), in Russian.
11. H.S. Carslaw, J.C. Jaeger, *Conduction of Heat in Solids*, Oxford University Press, London (1959).
12. D.E. Cagliostro, S.R. Riccitiello, K.J. Clark, A.B. Shimuzu, *J. Fire and Flammability*, vol. 205, 221 (1975).
13. J. Buckmaster, C. Anderseon, A. Nachman, *Internat. J. Eng. Sci.*, vol. 24, 263 (1986).
14. A.M. Kanury, D.J. Holve, *J. Heat Transfer*, vol. 104, 338 (1982).
15. A.M. Bulgakov, V.I. Kodolov, A.M. Lypanov, *Combustion Modelation of Polymer Materials, Chemistry*, Moscow (1990), in Russian.

16. R.M. Aseeva, G.E. Zaikov, *Combustion of Polymer Materials,* Hanser, Munich (1986).
17. T. Kashiwagi, 1987, Fall Technical Meeting Eastern States, Section of the Combustion Instituet, Nove. 26 (1987), NBS, Geithersberg, M.D.
18. N.A. Khalturinsky, Al.Al. Berlin. *Degradation and Stabilization of Polymers* by Ed. H.H.G. Jellinek, Elsevier, Amsterdam, vol. 2, ch. 3, 145 (1989).
19. R.M. Aseeva, G.E. Zaikov, *Developments in Polymer Stabilization* by Ed. G. Scott, Elsevier, App.l. Sci. Publ., London, vol. 7, Ch. 5., 233, (1984).
20. L.A. Lovachev, Academy Sciences News, USSR, ser. Chem, NI, 220 (1980).
21. C.J. Quinn, G.H. Beall. *Fourth Annual BCC COnference on Flame Retardancy: Recent Advances in Flame Retardancy of Polymeric Materials,* Stamford, Connecticut, May 18-20 (1993).
22. J. Green, *J. Fire Sci.,* vol. 10, Nov/Dec, 470 (1992); *Thermoplastic Polymer Additives/Theory and Practice* by Ed. J.T. Lutz, Marcel Deccer, NY, ch. 4, 93 (1989).
23. P. Hartey Des. Eng., Oct., 81 (1972).
24. Pat. USA 3878167.
25. Pat. USA 3874889.
26. Japan Pat. 36246, cl. 25 (I)A, 261 (CO8R).
27. Pat. USA 389577.
28. B.N. Dolgov, *Catalysis in Organical Chemistry,* Chemistry Publ., Moscow (1949), in Russian.
29. G. Camino, L. Costa, M.. Luda, *Fourth Annual BCC COnference on Flame Retardancy: Recent Advances in Flame Retardancy of Polymeric Materials;,* Stamford, Connecticut, May 18-20 (1993).
30. A.A. Dontsov, G.A. Lozovik, S.P. Novytskaya, *Chlorinated Polymers, Chemistry,* Moscow (1979), in Russian.
31. G.M. Ronkin, *Plastmassy,* N8, 16 (1980).
32. A.S. Levin, *Chlorinated Polyethylene and its Application,* by Ed. A.M. Savransy, Scientific Research Institute of Technical-Economic Investigations for Textile Industry, Moscow (1973).
33. V.I. Abramov, A.A. Krasheninnikova, Industry of chlorinated Polymers abroad, Sci. Research Inst. Techn. - *Econ. Invest. for Chemistry,* Moscow (1978).
34. Pat. USA 2640048.
35. Pat. USSR 583139 (1976).
36. Pat. USA 3347835.
37. Pat. USSR 364634 (1970).
38. Pat. USSR 547456 (1974).
39. Pat. Canada 471037.
40. *Chemie et Industrie* vol. 51, N5, 452 (1969).

41. Japan at. 27605 (1968).
42. *Industrial Chloro-organical Products*, by Ed. L.A. Oshin, Chemistry, Moscow (1978).
43. *Chem. Rundschau*, vol. 20, N9, 145, 147, 149 (1967).
44. *Rubber Age*, vol. 100, N10, 47 (1968).
45. *Express Inform. Synthetic High Molecular Materials*, N2, 13 (1976), in Russian.
46. *Polyethylene sulphochlorinated Technical Condition* N6-01-715-80, Ministry of Chem. Industry (1980), in Russian.
47. G.M. Ronkin, M.A. Korotyansky, A.I. Gershenovich, R.V. Dzhagatspanyan. *Caoutchouc and Rubber* N1, 5 (1980), in Russian.
48. G.M. Ronkin, M.A. Korotyansky, E.S Balakyrev, R.V. Dzhagatspanyan. *Plastmassy*, N3, 26 (1980).
49. O.V. Startsev, Yu.M. Vapyrov, A.S. Ovanesov, A.A. Donskoy, M.A. Shashkina. *High Molecular Compounds*, vol. 29A, N12, 2473 (1987).
50. *Caoutchouc and rubber*, N1, 5 (1980), in Russian.
51. G.M. Ronkin, Chlorosulphonated Polyethylene, Sci. Inv. Inst. Techn. Econom. *Inf. for Oil Chemistry Industry*, Moscow (1977).
52. V.G. Dyunina, *Investigation in area of Production porous sole Material based on chlorosulphonated Polyethylene*, Dissertation Textile Institute, Moscow (1972).
53. V.G. Dyunina, S.A. Pavlov, B.A. *Dinzburg Leather-Shoes Industry*, N5, 115 (1970), in Russian.
54. G.A. Blokh. *Organic Accelerators for vulcanization of caoutchoucs, Chemistry*, Moscow (1964), in Russian.
55. G.A. Blokh, *Organic Accelerators of culcanization and vulcanization systems for Elastomers*, Chemistry, Leningrad (1978).
56. I.I. Tugov, G.I. Kostyrina, *Chemistry and Physics of Polymers*, Chemistry, Moscow (1989), p. 431, in Russian.
57. A.A. Dontsov, S.P. Novytskaya, I.M. Kochanov, L.N. Stepanova, *Caoutchouc and Rubber* N3, 30 (1976).
58. V.S. Kuzin, A.A. Dontsov, G.M. Ronkin, M.A. Korotyansky ibid N11, 14 (1975).
59. A.A. Dontsov, *High Molecular Compounds*, vol. 18A, N1, 169 (1976).
60. A.A. Dontsov, S.P. Novytskaya, L.N. Stepanova, *Caoutchouc and Rubber*, N4, 24 (1977).
61. A.F. Nospycov, G.A. Blokh, ibid. N12, 18 (1983).
62. Pat. USSR 992539,   C08L 29/34.
63. Z.A. Kovacheva, G.A. Zhuravleva, R.G. Komyadyna, N.D. Trufanova, *Caoutchouc and Rubber* N11, 19 (1983).
64. L.T. Goncharova, A.G. Shvarts, V.A. Sapronov, ibid, N9, 18 91983).
65. Elastomer-Berichte-72, Du Pont Elastomer Hypalon. LD-974-10 Verzogerer Aktivator fur Hypalon; Chemische Bestendigkeit von Hypalon Vernetzungsprodukten.

66. A.L. Laabutin. *Caoutchoucs in anticorrosive technios,* Chemistry Publ., Moscow (1962).
67. Pat. USA 3883615.
68. Japan Pat. 57-129682, CO8L 23/08; declar 59-20341 (1982).
69. Japan decl. 59-18744 (1982), CO8L 23/06.
70. Japan decl. 40-14803 (1965).
71. FRG decl. 3042089 (1981).
72. France decl. 2419957 (1979), CO8L 23/28.
73. Japan decl. 54-26258 (1979).
74. A.A. Donskoy, M.A. Shashkina, R.M. Aseeva, L.V. Ruban, *Intern. J. Polym. Materials,* vol. 24, N1-4, 157 (1994).
75. R.M. Aseeva, L.V. Ruban, S.Kh. Korotkevich, A.A. Molchanov, G.E. Zaikov, *Plastmassy* N8, 78 (1989), in Russian.
76. V.V. Bogdanova, I.A. Klymovtsova, B.O. Phylonov, S.S. Phedeyev, a.o.. *High molecular compounds* vol. 28, B, N1, 42 (1987), in Russian.
77. S.S. Phedeyev, V.V. Svyrydov, V.V. Bogdanova, A.Ph. Surteyev a.o. Reports of Byelarus. Academy of Sciences, vol. 27, N1, 56 (1983).
78. L. Costa, G. Camino et al. *Polymer Degrad. and Stability* vol. 30, 13 (1990).
79. V.V. Bogdanova, I.A. Klymovtsova, S.S. Fedeev, A.I. Lesnikovich. *Intern. J. Polymer Materials,* vol. 2, 51 (1993).
80. J.W. Hastie, *Combustion and Flame,* vol. 77A, N6, 733 (1973).

SUBJECT INDEX